FOUNDATIONS
of ANALYSIS

FOUNDATIONS *of* ANALYSIS

STEVEN G. KRANTZ

Washington University
St. Louis, Missouri, USA

CRC Press
Taylor & Francis Group
Boca Raton London New York

CRC Press is an imprint of the
Taylor & Francis Group, an **informa** business
A CHAPMAN & HALL BOOK

CRC Press
Taylor & Francis Group
6000 Broken Sound Parkway NW, Suite 300
Boca Raton, FL 33487-2742

First issued in paperback 2019

ISBN-13: 978-1-4822-2074-2 (hbk)
ISBN-13: 978-1-138-37492-8 (pbk)

Visit the Taylor & Francis Web site at
http://www.taylorandfrancis.com

and the CRC Press Web site at
http://www.crcpress.com

To the memory of Bernhard Riemann, one of the fathers of modern analysis.

Table of Contents

Preface   ix

1   **Number Systems**   **1**
   1.1   The Real Numbers 1
   APPENDIX: Construction of the Real Numbers 6
   1.2   The Complex Numbers 11

2   **Sequences**   **19**
   2.1   Convergence of Sequences 19
   2.2   Subsequences 27
   2.3   Lim sup and Lim inf 31
   2.4   Some Special Sequences 35

3   **Series of Numbers**   **41**
   3.1   Convergence of Series 41
   3.2   Elementary Convergence Tests 48
   3.3   Advanced Convergence Tests 57
   3.4   Some Special Series 65
   3.5   Operations on Series 69

4   **Basic Topology**   **75**
   4.1   Open and Closed Sets 75
   4.2   Further Properties of Open and Closed Sets 82
   4.3   Compact Sets 87
   4.4   The Cantor Set 90
   4.5   Connected and Disconnected Sets 95
   4.6   Perfect Sets 98

5 Limits and Continuity of Functions   **101**
  5.1 Basic Properties of the Limit of a Function 101
  5.2 Continuous Functions 108
  5.3 Topological Properties and Continuity 113
  5.4 Classifying Discontinuities and Monotonicity 120

6 Differentiation of Functions   **127**
  6.1 The Concept of Derivative 127
  6.2 The Mean Value Theorem and Applications 134
  6.3 More on the Theory of Differentiation 142
  APPENDIX: Proof of Theorem 6.7 (The Weierstrass Nowhere Differentiable Function) 148

7 The Integral   **151**
  7.1 Partitions and the Concept of Integral 151
  7.2 Properties of the Riemann Integral 158

8 Sequences and Series of Functions   **169**
  8.1 Convergence of a Sequence of Functions 169
  8.2 More on Uniform Convergence 175
  8.3 Series of Functions 180
  8.4 The Weierstrass Approximation Theorem 185
  APPENDIX: Proof of the Weierstrass Approximation Theorem . . . 189

9 Elementary Transcendental Functions   **193**
  9.1 Power Series 193
  9.2 More on Power Series: Convergence Issues 200
  9.3 The Exponential and Trigonometric Functions 206
  9.4 Logarithms and Powers of Real Numbers 213

  APPENDIX I: Elementary Number Systems 217
  APPENDIX II: Logic and Set Theory 235
  Table of Notation 273
  Glossary 277
  Bibliography 293
  Index 295

Preface

Real analysis is a central part of modern mathematics. It is essential to most applications of mathematics—ranging from engineering to physics and beyond. Built on the basic ideas of the calculus, real analysis is the theoretical framework that holds everything together.

But real analysis, developed in its full armor, is a deep and difficult subject. It generally requires a full year course to cover all the basics. When I teach the course, I tell my students that this is the hardest course I ever took. And it will probably be the hardest course that they ever take.

The assertions in the last paragraph fly in the face of the fact that there are many end users of real analysis who simply do not need the full expertise and context that the subject has to offer. They need to understand the key ideas, and know how to apply the theorems, but they need not be masters of all the techniques. Thus there is need for a textbook that covers the basics of real analysis in a one-semester course. This is such a text.

How do we achieve these goals? First of all, we trim down the rigorous detail in the presentation. We do *not* present all the proofs. We do *not* require the students to learn all the proofs.

And we also have some didactic devices to make the path a little smoother. Each section begins with a boxed introduction called "Preliminary Remarks." These orient the student to the topics about to be presented, and set the stage for the work to be done. Each section ends with a blurb called "A Look Back." It gives the students about four questions to ponder with a view to reviewing what they have just learned. Also the text is peppered with enunciations entitled "Point of Confusion." These items note places where different pieces of terminology may seem to conflict with each other, or different ideas may not fit together properly. Finally, we have many "Remarks" to help put the material just presented in perspective.

A very important part of any real analysis text is the exercise sets. Each chapter generally has at least fifty exercises. We have an exercise set at the end of every section. This allows students to more easily key the exercises to the material in the section. And we are careful to stepladder the exercises. There is plenty of drill. The few challenging exercises are at the *end* of each problem set, and are marked with a *. So the exercises are a powerful and effective learning tool.

On the one hand, we want to provide the reader with needed background and review material—so that he she is not forced to run to the library to look things up. On the other hand we want to keep things streamlined, so that we get directly to the heart of the matter as quickly as possible. For this reason we put our review of number systems in Appendix I and of logic and proofs in Appendix II. Students who do not need this review can skip the Appendices. But many students will find themselves dipping into the Appendices just to look something up.

Generally speaking, the presentation in this text is smooth, concise, and elegant. We have made great efforts not to get bogged down in technicalities. Our main goal is to convey the key ideas in the most palatable possible fashion.

I thank my editor Robert Ross for his encouragement and guidance during this project. I am happy to express my appreciation to the reviewers engaged by the publisher to help bring this book to its current polished form; they contributed many interesting and useful ideas. I look forward to feedback from readers so that future editions can be made even more precise and accurate.

— Steven G. Krantz
St. Louis, Missouri

Chapter 1

Number Systems

1.1 The Real Numbers

Preliminary Remarks

Of course real analysis is about the real numbers. The real numbers are a very subtle number system with many deep and mysterious properties. In order to lay a proper foundation for our study of real analysis, we must first say precisely what the real numbers are and what their key properties are. That is our goal in this section.

This is a book about analysis in the real number system. Such a study must be founded on a careful consideration of *what the real numbers are* and *how they are constructed*. In the present section we give a careful treatment of the real number system. In the next we consider the complex numbers.

We know from calculus that, for many purposes, the rational numbers are inadequate. It is important to work in a number system which is closed with respect to the operations we shall perform—meaning that when you perform the operations you stay inside the set. For example, the integers are closed under addition because, when you add two integers, the answer you get is another integer. But the rational numbers are *not* closed under the operation of taking square root. Because the square root of 2 (the number $\sqrt{2}$) is *not* rational (see Appendix II).

This idea of closure includes limiting operations. While the rationals are closed under the usual arithmetic operations, they are not closed under the mathematical operation of *taking limits*. For instance, the sequence of rational numbers $3, 3.1, 3.14, 3.141, \ldots$ consists of terms that seem to be getting closer and closer together, *seem* to tend to some limit, and yet there is no rational number which will serve as a limit (of course it turns out that the limit is π—an "irrational" number).

We will now deal with the real number system, a system which contains all limits of sequences of rational numbers (as well as all limits of sequences of real numbers!). In fact our plan will be as follows: in this section we shall discuss all the requisite properties of the reals. The actual construction of the reals is rather complicated, and we shall put that in an Appendix to Section 1.1.

Definition 1.1 Let A be an ordered set and X a subset of A. The set X is called *bounded above* if there is an element $b \in A$ such that $x \leq b$ for all $x \in X$. We call the element b an *upper bound* for the set X.

EXAMPLE 1.2 Let $A = \mathbb{Q}$ (the rational numbers) with the usual ordering. The set $X = \{x \in \mathbb{Q} : 2 < x < 4\}$ is bounded above. For example, 15 is an upper bound for X. So are the numbers 12 and 4. It is interesting to observe that no element of this particular X can actually be an upper bound for X. The number 4 is a good candidate, but 4 is not an element of X. In fact if $b \in X$ then $(b+4)/2 \in X$ and $b < (b+4)/2 < 4$, so b could not be an upper bound for X. We think of 4 as a "least upper bound" for X (see below for a more thorough treatment of this idea). ∎

It turns out that the most convenient way to formulate the notion that the real numbers have "no holes" (i.e. that all sequences which seem to be converging actually have something to converge to) is in terms of upper bounds.

Definition 1.3 Let A be an ordered set and X a subset of A. An element $b \in A$ is called a *least upper bound* (or *supremum*) for X if b is an upper bound for X and $b \leq b'$ for every upper bound b' for X. We denote the supremum of X by $\sup X$. The supremum is also sometimes called the *least upper bound* and denoted by $\operatorname{lub} X$.

We shall see soon that the characterizing property of the real numbers is that every set with an upper bound has a least upper bound.

POINT OF CONFUSION 1.4 By its very definition, if a least upper bound exists then it is unique. The least upper bound is a tricky idea. Consider the set

$$\{x \in \mathbb{Q} : x^2 < 2\}.$$

This set is clearly bounded above—by 2 for example. So it has a least upper bound *in the real number system*. It does *not* have a least upper bound in the rational number system. Of course that least upper bound is $\sqrt{2}$.

EXAMPLE 1.5 In the last example, we considered the set X of rational numbers strictly between 2 and 4. We observed there that 4 is the least upper bound for X. Note that this least upper bound is not an element of the set X.

The set $Y = \{y \in \mathbb{Z} : -9 \leq y \leq 7\}$ has least upper bound 7. In this case, the least upper bound *is* an element of the set Y. ∎

Notice that we may define a lower bound for a subset of an ordered set in a fashion similar to that for an upper bound: $\ell \in A$ is a lower bound for $X \subseteq A$ if $\ell \leq x$ for all $x \in X$. An element $\ell \in A$ is called a *greatest lower bound* (or *infimum*) for X if ℓ is a lower bound for X and $\ell' \leq \ell$ for every lower bound ℓ' for X. We denote the infimum of X by $\inf X$. The infimum is also sometimes called the *greatest lower bound* and denoted by $\operatorname{glb} X$.

EXAMPLE 1.6 The set $X = \{x \in \mathbb{Q} : 2 < x < 4\}$ in the last two examples has lower bounds -20, 0, 1, 2, for instance. The greatest lower bound is 2, which is *not* an element of the set.

The set $Y = \{y \in \mathbb{Z} : -9 \leq y \leq 7\}$ in the last example has lower bounds—among others—given by $-53, -22, -10, -9$. The number -9 is the greatest lower bound. It *is* an element of Y. ∎

The purpose that the real numbers will serve for us is as follows: they will contain the rationals, they will still be an ordered field (that is, a set with operations of multiplication and addition and ordering which have reasonable properties such as commutativity and associativity—see the Appendix to this section for the details. Refer to [KRA1] for a thorough treatment of the concept of ordered field.). Also *every subset which has an upper bound will have a least upper bound*. We formulate this result as a theorem.

Theorem 1.7 *There exists an ordered field* $\mathbb{R}$ *which* (**i**) *contains* $\mathbb{Q}$ *and* (**ii**) *has the property that any nonempty subset of* $\mathbb{R}$ *which has an upper bound has a least upper bound (in the number system* $\mathbb{R}$).

The last property described in this theorem is called the Least Upper Bound Property of the real numbers. As mentioned previously, this theorem will be proved in the Appendix to Section 1.1. Now we begin to realize why it is so important to *construct* the number systems that we will use. We are endowing $\mathbb{R}$ with a great many properties. Why do we have any right to suppose that there exists a set with all these properties? We must produce one! We do so in the Appendix to Section 1.1.

Let us begin to explore the richness of the real numbers. The next theorem states a property which is certainly not shared by the rationals. It is fundamental in its importance.

Theorem 1.8 *Let x be a real number such that $x > 0$. Then there is a positive real number y such that $y^2 = y \cdot y = x$.*

Proof: We will use throughout this proof the fact that if $0 < a < b$ then $a^2 < b^2$.

Let
$$S = \{s \in \mathbb{R} : s > 0 \text{ and } s^2 < x\}.$$

Then S is not empty since $x/2 \in S$ if $x < 2$ and $1 \in S$ otherwise. Also S is bounded above since $x + 1$ is an upper bound for S. By Theorem 1.7, the set S has a least upper bound. Call it y. Obviously, $0 < \min\{x/2, 1\} \leq y$ hence y is positive. We claim that $y^2 = x$. To see this, we eliminate the other two possibilities.

If $y^2 < x$ then set $\epsilon = (x - y^2)/[4(x + 1)]$. Then $\epsilon > 0$ and

$$
\begin{aligned}
(y + \epsilon)^2 &= y^2 + 2 \cdot y \cdot \epsilon + \epsilon^2 \\
&= y^2 + 2 \cdot y \cdot \frac{x - y^2}{4(x + 1)} + \frac{x - y^2}{4(x + 1)} \cdot \frac{x - y^2}{4(x + 1)} \\
&< y^2 + 2 \cdot y \cdot \frac{x - y^2}{4y} + \frac{x - y^2}{4(x + 1)} \cdot \frac{x - y^2}{4(x + 1)} \\
&< y^2 + \frac{x - y^2}{2} + \frac{x - y^2}{4} \cdot \frac{x}{4x} \\
&< y^2 + (x - y^2) \\
&= x.
\end{aligned}
$$

Thus $y + \epsilon \in S$, and y cannot be an upper bound for S. This contradiction tells us that $y^2 \not< x$.

Similarly, if it were the case that $y^2 > x$ then we set $\epsilon = (y^2 - x)/[4(x + 1)]$. A calculation like the one we just did (see Exercise **2**) then shows that $(y - \epsilon)^2 \geq x$. Hence $y - \epsilon$ is also an upper bound for S, and y is therefore not the *least* upper bound. This contradiction shows that $y^2 \not> x$.

The only remaining possibility is that $y^2 = x$.   □

POINT OF CONFUSION 1.9 This last proof was fairly tricky. But it proves a very important fact—that every positive real number has a square root. This is in stark contrast to the situation for the rationals. We know, thanks to Pythagoras (see the details in Appendix II), that the rational number 2 does *not* have a square root *in the rationals*. But it certainly has a square root in the reals.

A similar proof shows that, if n is a positive integer and x a positive real number, then there is a positive real number y such that $y^n = x$. Exercise **15** asks you to provide the details.

We next use the Least Upper Bound Property of the Real Numbers to establish two important qualitative properties of the real numbers:

Theorem 1.10 *The set $\mathbb{R}$ of real numbers satisfies the Archimedean Property:*

> *Let a and b be positive real numbers. Then there is a natural number n such that $na > b$.*

The set $\mathbb{Q}$ of rational numbers satisfies the following Density Property:

> *Let $c < d$ be real numbers. Then there is a rational number q with $c < q < d$.*

Proof: Suppose the Archimedean Property to be false. Then $S = \{na : n \in \mathbb{N}\}$ has b as an upper bound. Therefore S has a finite supremum β. Since $a > 0$, it follows that $\beta - a < \beta$. So $\beta - a$ is not an upper bound for S, and there must be a natural number n' such that $n' \cdot a > \beta - a$. But then $(n'+1)a > \beta$, and β cannot be the supremum for S. This contradiction proves the first assertion.

For the second property, let $\lambda = d - c > 0$. By the Archimedean Property, choose a positive integer N such that $N \cdot \lambda > 1$. Again the Archimedean Property gives a natural number P such that $P > N \cdot c$ and another Q such that $Q > -N \cdot c$. Thus we see that Nc falls between the integers $-Q$ and P; therefore there must be an integer M between $-Q$ and P such that

$$M - 1 \leq Nc < M.$$

Thus $c < M/N$. Also

$$M \leq Nc + 1 \quad \text{hence} \quad \frac{M}{N} \leq c + \frac{1}{N} < c + \lambda = d.$$

So M/N is a rational number lying between c and d. $\qquad\square$

In Appendix II at the end of the book we establish that the set of all decimal representations of numbers is uncountable (see A2.7 in Appendix II). It follows that the set of all real numbers is uncountable. In fact the same proof shows that the set of all real numbers in the interval $(0, 1)$, or in any nonempty open interval (c, d), is uncountable.

POINT OF CONFUSION 1.11 Between every pair of distinct rational numbers there is an irrational number. And between every pair of distinct irrational numbers there is a rational number. But there are many more irrationals than rationals. So this is a very strange and complicated situation. The rationals and irrationals do *not* alternate. They are arranged in a complex and nonobvious fashion.

The set $\mathbb{R}$ of real numbers is uncountable, yet the set $\mathbb{Q}$ of rational numbers is countable. It follows that the set $\mathbb{R} \setminus \mathbb{Q}$ of *irrational* numbers is uncountable. In particular, it is nonempty. Thus we may see with very little effort that there exist a great many real numbers which cannot be expressed as a quotient of integers. However, it can be quite difficult to see whether any particular real number (such as π or e or $\sqrt[5]{2}$) is irrational.

We conclude by recalling the "absolute value" notation:

Definition 1.12 Let x be a real number. We define

$$|x| = \begin{cases} x & \text{if} & x > 0 \\ 0 & \text{if} & x = 0 \\ -x & \text{if} & x < 0 \end{cases}$$

It is left as an exercise for you to verify the important *triangle inequality*:

$$|x + y| \leq |x| + |y| \,.$$

[**Hint:** It is convenient to verify that the square of the lefthand side is less than or equal to the square of the righthand side.]

APPENDIX: Construction of the Real Numbers

There are several techniques for constructing the real number system $\mathbb{R}$ from the rational numbers system $\mathbb{Q}$. We use the method of Dedekind (Julius W. R. Dedekind, 1831-1916) cuts because it uses a minimum of new ideas and is fairly brief.

The number system that we shall be constructing is an instance of a *field* (the complex numbers, in the next section, also form a field). The definition is as follows:

Definition 1.13 A set S is called a *field* if it is equipped with a binary operation (usually called addition and denoted "+") and a second binary operation (called multiplication and denoted "·") such that the following axioms are satisfied (Here A stands for "addition," M stands for "multiplication," and D stands for "distributive law."):

A1. S is closed under addition: if $x, y \in S$ then $x + y \in S$.

A2. Addition is commutative: if $x, y \in S$ then $x + y = y + x$.

A3. Addition is associative: if $x, y, z \in S$ then $x + (y + z) = (x + y) + z$.

A4. There exists an element, called 0, in S which is an additive identity: if $x \in S$ then $0 + x = x$.

A5. Each element of S has an additive inverse: if $x \in S$ then there is an element $-x \in S$ such that $x + (-x) = 0$.

M1. S is closed under multiplication: if $x, y \in S$ then $x \cdot y \in S$.

M2. Multiplication is commutative: if $x, y \in S$ then $x \cdot y = y \cdot x$.

M3. Multiplication is associative: if $x, y, z \in S$ then $x \cdot (y \cdot z) = (x \cdot y) \cdot z$.

M4. There exists an element, called 1, which is a multiplicative identity: if $x \in S$ then $x \cdot 1 = x$.

M5. Each nonzero element of S has a multiplicative inverse: if $0 \neq x \in S$ then there is an element $x^{-1} \in S$ such that $x \cdot (x^{-1}) = 1$. The element x^{-1} is sometimes denoted $1/x$.

D1. Multiplication distributes over addition: if $x, y, z \in S$ then

$$x \cdot (y + z) = x \cdot y + x \cdot z.$$

Definition 1.14 A *cut* is a subset $\mathcal{P}$ of $\mathbb{Q}$ with the following properties:

- $\mathcal{P} \neq \emptyset$

- If $s \in \mathcal{P}$ and $t < s$ then $t \in \mathcal{P}$

- If $s \in \mathcal{P}$ then there is a $u \in \mathcal{P}$ such that $u > s$

- There is a rational number x such that $c < x$ for all $c \in \mathcal{P}$

You should think of a cut $\mathcal{P}$ as the set of all rational numbers to the left of some point in the real line. Since we have not constructed the real line yet, we cannot define a cut in that simple way; we have to make the construction more indirect. But if you consider the four properties of a cut, they describe a set that looks like a "rational halfline."

Notice that, if $\mathcal{P}$ is a cut and the point $s \notin \mathcal{P}$, then any rational $t > s$ is also not in $\mathcal{P}$. Also, if $r \in \mathcal{P}$ and $s \notin \mathcal{P}$ then it must be that $s > r$.

Definition 1.15 If $\mathcal{P}$ and $\mathcal{D}$ are cuts then we say that $\mathcal{P} < \mathcal{D}$ provided that $\mathcal{P}$ is a subset of $\mathcal{D}$ but $\mathcal{P} \neq \mathcal{D}$.

Check for yourself that "<" is an ordering on the set of all cuts.

Now we introduce operations of addition and multiplication which will turn the set of all cuts into a field.

Definition 1.16 If $\mathcal{P}$ and $\mathcal{D}$ are cuts then we define

$$\mathcal{P} + \mathcal{D} = \{c + d : c \in \mathcal{P}, d \in \mathcal{D}\}.$$

We define the cut $\widehat{0}$ to be the set of all negative rationals.

The cut $\widehat{0}$ will play the role of the additive identity. We are now required to check that field axioms **A1-A5** hold.

For **A1**, we need to see that $\mathcal{P} + \mathcal{D}$ is a cut. Obviously $\mathcal{P} + \mathcal{D}$ is not empty. If s is an element of $\mathcal{P} + \mathcal{D}$ and t is a rational number less than s, write $s = c + d$, where $c \in \mathcal{P}$ and $d \in \mathcal{D}$. Then $t - c < s - c = d \in \mathcal{D}$ so $t - c \in \mathcal{D}$; and $c \in \mathcal{P}$. Hence $t = c + (t - c) \in \mathcal{P} + \mathcal{D}$. A similar argument shows that there is an $r > s$ such that $r \in \mathcal{P} + \mathcal{D}$. Finally, if x is a rational upper bound for $\mathcal{P}$ and y is a rational upper bound for $\mathcal{D}$, then $x + y$ is a rational upper bound for $\mathcal{P} + \mathcal{D}$. We conclude that $\mathcal{P} + \mathcal{D}$ is a cut.

Since addition of rational numbers is commutative, it follows immediately that addition of cuts is commutative. Associativity follows in a similar fashion.

Now we show that, if $\mathcal{P}$ is a cut, then $\mathcal{P} + \widehat{0} = \mathcal{P}$. For if $c \in \mathcal{P}$ and $z \in \widehat{0}$ then $c + z < c + 0 = c$ hence $\mathcal{P} + \widehat{0} \subseteq \mathcal{P}$. Also, if $c' \in \mathcal{P}$ then choose a $d' \in \mathcal{P}$ such that $c' < d'$. Then $c' - d' < 0$ so $c' - d' \in \widehat{0}$. And $c' = d' + (c' - d')$. Hence $\mathcal{P} \subseteq \mathcal{P} + \widehat{0}$. We conclude that $\mathcal{P} + \widehat{0} = \mathcal{P}$.

Finally, for Axiom **A5**, we let $\mathcal{P}$ be a cut and set $-\mathcal{P}$ to be equal to $\{d \in \mathbb{Q} : c + d < 0 \text{ for all } c \in \mathcal{P}\}$. If x is a rational upper bound for $\mathcal{P}$ and $c \in \mathcal{P}$ then $-x \in -\mathcal{P}$ so $-\mathcal{P}$ is not empty. By its very definition, $\mathcal{P} + (-\mathcal{P}) \subseteq \widehat{0}$. Further, if $z \in \widehat{0}$ and $c \in \mathcal{P}$ we set $c' = z - c$. Then $c' \in -\mathcal{P}$ and $z = c + c'$. Hence $\widehat{0} \subseteq \mathcal{P} + (-\mathcal{P})$. We conclude that $\mathcal{P} + (-\mathcal{P}) = \widehat{0}$.

Having verified the axioms for addition, we turn now to multiplication.

Definition 1.17 If $\mathcal{P}$ and $\mathcal{D}$ are cuts then we define the product $\mathcal{P} \cdot \mathcal{D}$ as follows:

- If $\mathcal{P}, \mathcal{D} > \widehat{0}$ then $\mathcal{P} \cdot \mathcal{D} = \{q \in \mathbb{Q} : q < c \cdot d \text{ for some } c \in \mathcal{P}, d \in \mathcal{D} \text{ with } c > 0, d > 0 \}$

- If $\mathcal{P} > \widehat{0}, \mathcal{D} < \widehat{0}$ then $\mathcal{P} \cdot \mathcal{D} = -(\mathcal{P} \cdot (-\mathcal{D}))$

- If $\mathcal{P} < \widehat{0}, \mathcal{D} > \widehat{0}$ then $\mathcal{P} \cdot \mathcal{D} = -((-\mathcal{P}) \cdot \mathcal{D})$

- If $\mathcal{P}, \mathcal{D} < \widehat{0}$ then $\mathcal{P} \cdot \mathcal{D} = (-\mathcal{P}) \cdot (-\mathcal{D})$

- If either $\mathcal{P} = \widehat{0}$ or $\mathcal{D} = \widehat{0}$ then $\mathcal{P} \cdot \mathcal{D} = \widehat{0}$.

Notice that, for convenience, we have defined multiplication of negative numbers just as we did in high school. The reason is that the definition that we use for the product of two positive numbers cannot work when one of the two factors is negative (exercise).

It is now a routine exercise to verify that the set of all cuts, with this definition of multiplication, satisfies field axioms **M1-M5**. The proofs follow those for **A1-A5** rather closely.

For the distributive property, one first checks the case when all the cuts are positive, reducing it to the distributive property for the rationals. Then one handles negative cuts on a case by case basis.

We now know that the collection of all cuts forms an ordered field. Denote this field by the symbol $\mathbb{R}$. We next verify the crucial property of $\mathbb{R}$ that sets it apart from $\mathbb{Q}$:

Theorem 1.18 *The ordered field $\mathbb{R}$ satisfies the Least Upper Bound Property.*

Proof: Let S be a subset of $\mathbb{R}$ which is bounded above. Define

$$\mathcal{S}^* = \bigcup_{\mathcal{P} \in S} \mathcal{P}.$$

Then $\mathcal{S}^*$ is clearly nonempty, and it is therefore a cut since it is a union of cuts. It is also clearly an upper bound for S since it contains each element of S. It remains to check that $\mathcal{S}^*$ is the least upper bound for S.

In fact if $\mathcal{T} < \mathcal{S}^*$ then $\mathcal{T} \subseteq \mathcal{S}^*$ and there is a rational number q in $\mathcal{S}^* \setminus \mathcal{T}$. But, by the definition of $\mathcal{S}^*$, it must be that $q \in \mathcal{P}$ for some $\mathcal{P} \in S$. So $\mathcal{P} > \mathcal{T}$, and $\mathcal{T}$ cannot be an upper bound for S. Therefore $\mathcal{S}^*$ is the least upper bound for S, as desired. $\square$

We have shown that $\mathbb{R}$ is an ordered field which satisfies the Least Upper Bound Property. It remains to show that $\mathbb{R}$ contains (a copy of) $\mathbb{Q}$ in a natural way. In fact, if $q \in \mathbb{Q}$ we associate to it the element $\varphi(q) = \mathcal{P}_q \equiv \{x \in \mathbb{Q} : x < q\}$. Then $\mathcal{P}_q$ is obviously a cut. It is also routine to check that

$$\varphi(q + q') = \varphi(q) + \varphi(q') \quad \text{and} \quad \varphi(q \cdot q') = \varphi(q) \cdot \varphi(q').$$

Therefore we see that φ represents $\mathbb{Q}$ as a subfield of $\mathbb{R}$.

A Look Back

1. What is a least upper bound?
2. Does the least upper bound of a set X necessarily lie in X?
3. What is the defining property of the real numbers?
4. How do the real numbers differ from the rational numbers?

Exercises

1. Let A be a set of real numbers that is bounded above and set $\alpha = \sup A$. Let $B = \{-a : a \in A\}$. Prove that $\inf B = -\alpha$. Prove the same result with the roles of infimum and supremum reversed.

2. Complete the calculation in the proof of Theorem 1.8.

3. What is the least upper bound of the set

$$S = \{x : x^2 < 2\}?$$

   Explain why this question has a sensible answer in the real number system but not in the rational number system.

4. Prove that the least upper bound and greatest lower bound for a set of real numbers are unique.

5. Consider the unit circle C. Let

$$S = \{\alpha : 2\alpha < \text{ (the circumference of } C)\}.$$

   Show that S is bounded above. Let p be the least upper bound of S. Say explicitly what the number p is. This exercise works in the real number system, but not in the rational number system. Why?

6. Give an example of a set that contains its least upper bound but not its greatest lower bound. Give an example of a set that contains its greatest lower bound but not its least upper bound.

7. Give an example of a set of real numbers that does *not* have a least upper bound. Give an example of a set of real numbers that does *not* have a greatest lower bound.

8. Prove the triangle inequality.

9. Prove that addition of the real numbers (as constructed in the Appendix to Section 1.1) is commutative. Now prove that it is associative.

10. Use the triangle inequality to prove that $|a-b| \leq |a|+|b|$ for any real numbers a and b.

11. Let $\emptyset$ be the empty set (see Appendix II). Prove that $\sup \emptyset = -\infty$ and $\inf \emptyset = +\infty$.

12. Use the triangle inequality to prove that $|a| - |b| \leq |a-b|$ for any real numbers a and b.

* 13. Let f be a function with domain the reals and range the reals. Assume that f has a local minimum at each point x in its domain. (This means that, for each $x \in \mathbb{R}$, there is an $\epsilon > 0$ such that whenever $|x-t| < \epsilon$ then $f(x) \leq f(t)$). *Do not assume that f is differentiable, or continuous, or anything nice like that.* Prove that the image of f is countable. (**Hint:** When this author solved this problem as a student, his solution was ten pages long; however, there is a one-line solution due to Michael Spivak.)

*   14. Let λ be a positive irrational real number. If n is a positive integer, choose
by the Archimedean Property an integer k such that $k\lambda \leq n < (k+1)\lambda$. Let
$\varphi(n) = n - k\lambda$. Prove that the set of all $\varphi(n)$ is dense in the interval $[0, \lambda]$.
[By this we mean that the $\varphi(n)$ get arbitrarily close to each element of $[0, \lambda]$.]
(**Hint:** Examine the proof of the density of the rationals in the reals.)

*   15. Let n be a natural number and x a positive real number. Prove that there is
a positive real number y such that $y^n = x$. Is y unique?

1.2   The Complex Numbers

Preliminary Remarks

In this book we do not emphasize the complex numbers. But we do use them occasionally, so this section is provided for your reference. Be sure that, at the least, you understand the basic arithmetic operations on $\mathbb{C}$ and also the special role of the complex number i.

When we first learn about the complex numbers, the most troublesome point is the very beginning: "Let's pretend that the number -1 has a square root. Call it i." What gives us the right to "pretend" in this fashion? The answer is that we have no such right.[1] If -1 has a square root, then we should be able to construct a number system in which that is the case. That is what we shall do in this section.

Definition 1.19 The system of *complex numbers*, denoted by the symbol $\mathbb{C}$, consists of all ordered pairs (a, b) of real numbers. We add two complex numbers (a, b) and $(\tilde{a}, \tilde{b})$ by the formula

$$(a, b) + (\tilde{a}, \tilde{b}) = (a + \tilde{a}, b + \tilde{b}).$$

We multiply two complex numbers by the formula

$$(a, b) \cdot (\tilde{a}, \tilde{b}) = (a \cdot \tilde{a} - b \cdot \tilde{b}, a \cdot \tilde{b} + \tilde{a} \cdot b).$$

---

[1] The complex numbers were initially developed so that we would have a number system in which all polynomial equations are solvable. One of the reasons, historically, that mathematicians had trouble accepting the complex numbers is that they did not believe that they really existed—they were just made up. This is, in part, how they came to be called "imaginary." Mathematicians had similar trouble accepting negative numbers; for a time, negative numbers were called "forbidden."

Remark 1.20 If you are puzzled by this definition of multiplication, do not worry. In a few moments you will see that it gives rise to the notion of multiplication of complex numbers that you are accustomed to. Perhaps more importantly, a naive rule for multiplication like $(a, b) \cdot (\tilde{a}, \tilde{b}) = (a\tilde{a}, b\tilde{b})$ gives rise to nonsense like $(1, 0) \cdot (0, 1) = (0, 0)$. It is really necessary for us to use the initially counterintuitive definition of multiplication that is presented here.

EXAMPLE 1.21 Let $z = (3, -2)$ and $w = (4, 7)$ be two complex numbers. Then

$$z + w = (3, -2) + (4, 7) = (3 + 4, -2 + 7) = (7, 5).$$

Also

$$z \cdot w = (3, -2) \cdot (4, 7) = (3 \cdot 4 - (-2) \cdot 7, 3 \cdot 7 + 4 \cdot (-2)) = (26, 13). \quad \blacksquare$$

As usual, we ought to check that addition and multiplication are commutative and associative, that multiplication distributes over addition, and so forth. We shall leave these tasks to the exercises. Instead we develop some of the crucial, and more interesting, properties of our new number system.

Theorem 1.22 *The following properties hold for the number system* $\mathbb{C}$.

(a) *The number* $1 \equiv (1, 0)$ *is the multiplicative identity:* $1 \cdot z = z$ *for any* $z \in \mathbb{C}$.

(b) *The number* $0 \equiv (0, 0)$ *is the additive identity:* $0 + z = z$ *for any* $z \in \mathbb{C}$.

(c) *Each complex number* $z = (x, y)$ *has an additive inverse* $-z = (-x, -y)$: *it holds that* $z + (-z) = 0$.

(d) *The number* $i \equiv (0, 1)$ *satisfies* $i \cdot i = -1$; *in other words,* i *is a square root of* -1.

Proof: These are direct calculations, but it is important for us to work out these facts.

First, let $z = (x, y)$ be any complex number. Then

$$1 \cdot z = (1, 0) \cdot (x, y) = (1 \cdot x - 0 \cdot y, 1 \cdot y + x \cdot 0) = (x, y) = z.$$

This proves the first assertion.

For the second, we have

$$0 + z = (0, 0) + (x, y) = (0 + x, 0 + y) = (x, y) = z.$$

With z as above, set $-z = (-x, -y)$. Then

$$z + (-z) = (x, y) + (-x, -y) = (x + (-x), y + (-y)) = (0, 0) = 0.$$

Finally, we calculate

$$i \cdot i = (0,1) \cdot (0,1) = (0 \cdot 0 - 1 \cdot 1, 0 \cdot 1 + 0 \cdot 1) = (-1,0) = -1 \,.$$

Thus, as asserted, i is a square root of -1. □

POINT OF CONFUSION 1.23 The model for the real numbers is the set of all Dedekind cuts. This is a non-obvious construction that guarantees that there really is a number system that satisfies the important and nontrivial properties of the reals.

By contrast, the complex numbers are very easy to construct. The complex number system is simply the two-dimensional Cartesian plane equipped with some interesting algebraic operations.

Proposition 1.24 *If $z \in \mathbb{C}, z \neq 0$, then there is a complex number w such that $z \cdot w = 1$.*

Proof: Write $z = (x,y)$ and set

$$w = \left(\frac{x}{\sqrt{x^2 + y^2}}, \frac{-y}{\sqrt{x^2 + y^2}} \right) \,.$$

Since $z \neq 0$, this definition makes sense. Then it is straightforward to verify that $z \cdot w = 1$. □

Thus every nonzero complex number has a multiplicative inverse. The other field axioms for $\mathbb{C}$ are easy to check. We conclude that the number system $\mathbb{C}$ forms a field. You will prove in the exercises that it is not possible to order this field. If α is a real number then we associate α with the complex number $(\alpha, 0)$. Thus we have the natural "embedding"

$$\mathbb{R} \ni \alpha \longmapsto (\alpha, 0) \in \mathbb{C} \,.$$

In this way, we can think of the real numbers as a *subset* of the complex numbers. In fact, the real field $\mathbb{R}$ is a *subfield* of the complex field $\mathbb{C}$. This means that if $\alpha, \beta \in \mathbb{R}$ and $(\alpha, 0), (\beta, 0)$ are the corresponding elements in $\mathbb{C}$ then $\alpha + \beta$ corresponds to $(\alpha + \beta, 0)$ and $\alpha \cdot \beta$ corresponds to $(\alpha, 0) \cdot (\beta, 0)$. These assertions are explored more thoroughly in the exercises.

With the remarks in the preceding paragraph we can sometimes ignore the distinction between the real numbers and the complex numbers. For example, we can write

$$5 \cdot i$$

and understand that it means $(5,0) \cdot (0,1) = (0,5)$. Likewise, the expression

$$5 \cdot 1$$

can be interpreted as $5 \cdot 1 = 5$ or as $(5,0) \cdot (1,0) = (5,0)$ without any danger of ambiguity.

Theorem 1.25 *Every complex number can be written in the form $a + b \cdot i$, where a and b are real numbers. In fact, if $z = (x,y) \in \mathbb{C}$ then*

$$z = x + y \cdot i \,.$$

Proof: With the identification of real numbers as a subfield of the complex numbers, we have that

$$x + y \cdot i = (x,0) + (y,0) \cdot (0,1) = (x,0) + (0,y) = (x,y) = z$$

as claimed.   □

Now that we have constructed the complex number field, we will adhere to the usual custom of writing complex numbers as $z = a+b{\cdot}i$ or, more simply, $a + bi$. We call a the *real part* of z, denoted by $\mathrm{Re}\, z$, and b the *imaginary part* of z, denoted $\mathrm{Im}\, z$. We have

$$(a + bi) + (\tilde{a} + \tilde{b}i) = (a + \tilde{a}) + (b + \tilde{b})i$$

and

$$(a + bi) \cdot (\tilde{a} + \tilde{b}i) = (a \cdot \tilde{a} - b \cdot \tilde{b}) + (a \cdot \tilde{b} + \tilde{a} \cdot b)i \,.$$

If $z = a + bi$ is a complex number then we define its *complex conjugate* to be the number $\bar{z} = a - bi$. We record some elementary facts about the complex conjugate:

Proposition 1.26 *If z, w are complex numbers then*

(1) $\overline{z + w} = \bar{z} + \bar{w}$;

(2) $\overline{z \cdot w} = \bar{z} \cdot \bar{w}$;

(3) $z + \bar{z} = 2 \cdot \mathrm{Re}\, z$;

(4) $z - \bar{z} = 2 \cdot i \cdot \mathrm{Im}\, z$;

(5) $z \cdot \bar{z} \geq 0$, with equality holding if and only if $z = 0$.

Proof: Write $z = a + bi$, $w = c + di$. Then

$$
\begin{aligned}
\overline{z + w} &= \overline{(a + c) + (b + d)i} \\
&= (a + c) - (b + d)i \\
&= (a - bi) + (c - di) \\
&= \bar{z} + \bar{w}.
\end{aligned}
$$

This proves **(1)**. Assertions **(2)**, **(3)**, **(4)** are proved similarly.
For **(5)**, notice that

$$z \cdot \overline{z} = (a + bi) \cdot (a - bi) = a^2 + b^2 \geq 0.$$

Clearly equality holds if and only if $a = b = 0$. □

POINT OF CONFUSION 1.27 The concept of complex conjugate was invented for the following reason. If p is a polynomial with real coefficients, and if α is a complex root of p (so $p(\alpha) = 0$), then $\overline{\alpha}$ is also a complex root of p. It is an exercise for you to confirm this statement.

The expression $|z|$ is defined to be the nonnegative square root of $z \cdot \overline{z}$:

$$|z| = +\sqrt{z \cdot \overline{z}} = \sqrt{x^2 + y^2}$$

when $z = x + iy$. It is called the *modulus* of z and plays the same role for the complex field that absolute value plays for the real field. It is the distance of z to the origin. The modulus has the following properties.

Proposition 1.28 *If $z, w \in \mathbb{C}$ then*

(1) $|z| = |\overline{z}|$;

(2) $|z \cdot w| = |z| \cdot |w|$;

(3) $|\mathrm{Re}\, z| \leq |z|$, $|\mathrm{Im}\, z| \leq |z|$;

(4) $|z + w| \leq |z| + |w|$;

Proof: Write $z = a + bi, w = c + di$. Then **(1)**, **(2)**, **(3)** are immediate. For **(4)** we calculate that

$$
\begin{aligned}
|z + w|^2 &= (z + w) \cdot (\overline{z + w}) \\
&= z \cdot \overline{z} + z \cdot \overline{w} + w \cdot \overline{z} + w \cdot \overline{w} \\
&= |z|^2 + 2\mathrm{Re}\,(z \cdot \overline{w}) + |w|^2 \\
&\leq |z|^2 + 2|z \cdot \overline{w}| + |w|^2 \\
&= |z|^2 + 2|z| \cdot |w| + |w|^2 \\
&= (|z| + |w|)^2.
\end{aligned}
$$

Taking square roots proves **(4)**. □

Observe that, if z is real, then $z = a + 0i$ and the modulus of z equals the absolute value of a. Likewise, if $z = 0 + bi$ is pure imaginary, then the modulus of z equals the absolute value of b. In particular, the fourth part of the proposition reduces, in the real case, to the triangle inequality

$$|a + b| \le |a| + |b|.$$

If z is any nonzero complex number, then let $r = |z|$. Now define $\xi = z/r$. Certainly ξ is a complex number of modulus 1. Thus ξ lies on the unit circle, so it subtends an angle θ with the positive x-axis. Certainly then $\xi = \cos\theta + i\sin\theta$. It is shown in Section 9.3 that

$$e^{i\theta} = \xi = \cos\theta + i\sin\theta.$$

[**Hint:** You may verify this formula for yourself by writing out the power series for the exponential and writing out the power series for cosine and sine.] As a result, we may write

$$z = re^{i\theta}.$$

We conclude this discussion by recording the most important basic fact about the complex numbers. Carl Friedrich Gauss (1777–1855) gave five proofs of this theorem (the Fundamental Theorem of Algebra) in his doctoral dissertation:

Theorem 1.29 *Let $p(z)$ be any polynomial of degree at least 1. Then p has a root $\alpha \in \mathbb{C}$ such that $p(\alpha) = 0$.*

POINT OF CONFUSION 1.30 Using a little algebra, one can in fact show that a polynomial of degree k has k roots (counting multiplicity).

For example, the polynomial $p(z) = z^4 - 2z^3 + 2z^2 - 2z + 1$ has roots $i, -i, 1$, and 1. Put in other words,

$$p(z) = (z - i)(z + i)(z - 1)(z - 1).$$

A Look Back

1. What is the definition of the complex numbers?

2. How do we multiply two complex numbers?

3. What is the special role of the complex number i?

4. Why is it the case that every nonzero complex number has two distinct square roots?

Exercises

1. Taking the commutative, associative, and distributive laws for the real number system for granted, establish these laws for the complex numbers.

2. Consider the function $\phi : \mathbb{R} \to \mathbb{C}$ given by $\phi(x) = x + i \cdot 0$. Prove that ϕ respects addition and multiplication in the sense that $\phi(x+x') = \phi(x)+\phi(x')$ and $\phi(x \cdot x') = \phi(x) \cdot \phi(x')$.

3. If $z, w \in \mathbb{C}$ then prove that $\overline{z/w} = \overline{z}/\overline{w}$.

4. Prove that the set of all complex numbers is uncountable.

5. Prove that the set of all complex numbers with rational real part is uncountable.

6. Prove that the set of all complex numbers with both real and imaginary parts rational is countable.

7. Prove that the set $\{z \in \mathbb{C} : |z| = 1\}$ is uncountable.

8. Prove that the field of complex numbers cannot be made into an *ordered* field. (**Hint:** Since $i \neq 0$ then either $i > 0$ or $i < 0$. Both lead to a contradiction.)

9. Find all cube roots of the complex number $1 + i$.

10. Use the Fundamental Theorem of Algebra to prove that any polynomial of degree k has k (not necessarily distinct) roots.

11. Prove that the complex roots of a polynomial with real coefficients occur in complex conjugate pairs.

12. Calculate the square roots of i.

13. In the complex plane, draw a picture of
$$S = \{z \in \mathbb{C} : |z - 1| + |z + 1| = 2\}.$$

14. In the complex plane, draw a picture of
$$T = \{z \in \mathbb{C} : |z + \overline{z}| - |z - \overline{z}| = 2\}.$$

15. Prove that any nonzero complex number has kth roots $r_1, r_2, \ldots, r_k$. That is, prove that there are k of them.

Chapter 2

Sequences

2.1 Convergence of Sequences

> ### Preliminary Remarks
>
> Sequences are the nuts and bolts of real analysis. All the basic ideas are formulated in terms of sequences. In particular, we care about *limits* of sequences. We need to be able to determine when a sequence has a limit, and we often need to calculate that limit. This section introduces you to these ideas.

A *sequence* of real numbers is a function $\varphi : \mathbb{N} \to \mathbb{R}$. We often write the sequence as $\varphi(1), \varphi(2), \ldots$ or, more simply, as $\varphi_1, \varphi_2, \ldots$.

EXAMPLE 2.1 The function $\varphi(j) = 1/j$ is a sequence of real numbers. We will often write such a sequence as $\varphi_j = 1/j$ or as $\{1, 1/2, 1/3, \ldots\}$ or as $\{1/j\}_{j=1}^{\infty}$. ∎

POINT OF CONFUSION 2.2 Do not be misled into thinking that a sequence must form a pattern, or be given by a formula. Obviously the ones which are given by formulas are easy to write down, but they are certainly not typical. For example, the coefficients in the decimal expansion of π, $\{3, 1, 4, 1, 5, 9, 2, \ldots\}$, fit our definition of sequence—but they are not given by any obvious pattern.

The most important question about a sequence is whether it converges. We define this notion as follows.

Definition 2.3 A sequence $\{a_j\}$ of real numbers is said to *converge* to a real number α if, for each $\epsilon > 0$, there is an integer $N > 0$ such that if $j > N$ then

$|a_j - \alpha| < \epsilon$. We call α the *limit* of the sequence $\{a_j\}$. We write $\lim_{j \to \infty} a_j = \alpha$. We also sometimes write $a_j \to \alpha$.

If a sequence $\{a_j\}$ does not converge then we frequently say that it *diverges*.

EXAMPLE 2.4 Let $a_j = 1/j, j = 1, 2, \ldots$. Then the sequence converges to 0. For let $\epsilon > 0$. Choose N to be the next integer after $1/\epsilon$. If $j > N$, then

$$|a_j - 0| = |a_j| = \frac{1}{j} < \frac{1}{N} < \epsilon,$$

proving the claim.

Let $b_j = (-1)^j, j = 1, 2, \ldots$. Then the sequence *does not converge*. To prove this assertion, suppose to the contrary that it does. Say that the sequence converges to a number α. Let $\epsilon = 1/2$. By definition of convergence, there is an integer $N > 0$ such that, if $j > N$, then $|b_j - \alpha| < \epsilon = 1/2$. For such j we have

$$2 = |(-1)^j - (-1)^{j+1}| = |b_j - b_{j+1}| \leq |b_j - \alpha| + |\alpha - b_{j+1}|$$

(by the triangle inequality—see the end of Section 1.1). But this last is

$$< \epsilon + \epsilon = 1.$$

We have proved that $2 < 1$, a clear contradiction. So the sequence $\{b_j\}$ has no limit.   ∎

POINT OF CONFUSION 2.5 Given any sequence, it either converges or it diverges. There is no in-between status, and no undecided status.

We begin with a few intuitively appealing properties of convergent sequences which will be needed later. First, a definition.

Definition 2.6 A sequence a_j is said to be *bounded* if there is a number $M > 0$ such that $|a_j| \leq M$ for every j.

Now we have

Proposition 2.7 *Let* $\{a_j\}$ *be a convergent sequence. Then we have:*

- *The limit of the sequence is unique.*

- *The sequence is bounded.*

Proof: Suppose that the sequence has two limits α and $\tilde{\alpha}$. Let $\epsilon > 0$. Then there is an integer $N > 0$ such that for $j > N$ we have the inequality $|a_j - \alpha| < \epsilon/2$. Likewise, there is an integer $\tilde{N} > 0$ such that for $j > \tilde{N}$ we have $|a_j - \tilde{\alpha}| < \epsilon/2$. Let $N_0 = \max\{N, \tilde{N}\}$. Then, for $j > N_0$, we have

$$|\alpha - \tilde{\alpha}| \leq |\alpha - a_j| + |a_j - \tilde{\alpha}| < \epsilon/2 + \epsilon/2 = \epsilon \,.$$

Since this inequality holds for any $\epsilon > 0$ we have that $\alpha = \tilde{\alpha}$. So the limit of the sequence is unique.

Next, with α the limit of the sequence and $\epsilon = 1$, we choose an integer $N > 0$ such that $j > N$ implies that $|a_j - \alpha| < \epsilon = 1$. For such j we have that

$$|a_j| \leq |a_j - \alpha| + |\alpha| < 1 + |\alpha| \equiv P \,.$$

Let $Q = \max\{|a_1|, |a_2|, \ldots, |a_N|\}$. If j is any natural number then either $1 \leq j \leq N$ (in which case $|a_j| \leq Q$) or else $j > N$ (in which case $|a_j| \leq P$). Set $M = \max\{P, Q\}$. Then $|a_j| \leq M$ for all j, as desired. So the sequence is bounded. $\qquad\square$

The next proposition records some elementary properties of limits of sequences.

Proposition 2.8 *Let $\{a_j\}$ be a sequence of real numbers with limit α and $\{b_j\}$ be a sequence of real numbers with limit β. Then we have:*

(1) *If c is a constant then the sequence $\{c \cdot a_j\}$ converges to $c \cdot \alpha$;*

(2) *The sequence $\{a_j + b_j\}$ converges to $\alpha + \beta$;*

(3) *The sequence $a_j \cdot b_j$ converges to $\alpha \cdot \beta$;*

(4) *If $b_j \neq 0$ for all j and $\beta \neq 0$ then the sequence a_j/b_j converges to α/β.*

Proof: For the first part, we may assume that $c \neq 0$ (for when $c = 0$ there is nothing to prove). Let $\epsilon > 0$. Choose an integer $N > 0$ such that for $j > N$ it holds that

$$|a_j - \alpha| < \frac{\epsilon}{|c|} \,.$$

For such j we have that

$$|c \cdot a_j - c \cdot \alpha| = |c| \cdot |a_j - \alpha| < |c| \cdot \frac{\epsilon}{|c|} = \epsilon \,.$$

This proves the first assertion.

The proof of the second assertion is similar, and we leave it as an exercise.

For the third assertion, notice that the sequence $\{a_j\}$ is bounded (by the second part of Proposition 2.7): say that $|a_j| \leq M$ for every j. Let $\epsilon > 0$. Choose an integer $N > 0$ so that $|a_j - \alpha| < \epsilon/(2M + 2|\beta|)$ when $j > N$. Also choose an integer $\widetilde{N} > 0$ such that $|b_j - \beta| < \epsilon/(2M + 2|\beta|)$ when $j > \widetilde{N}$. Then, for $j > \max\{N, \widetilde{N}\}$, we have that

$$
\begin{aligned}
|a_j b_j - \alpha\beta| &= |a_j(b_j - \beta) + \beta(a_j - \alpha)| \\
&\leq |a_j(b_j - \beta)| + |\beta(a_j - \alpha)| \\
&< M \cdot \frac{\epsilon}{2M + 2|\beta|} + |\beta| \cdot \frac{\epsilon}{2M + 2|\beta|} \\
&\leq \frac{\epsilon}{2} + \frac{\epsilon}{2} \\
&= \epsilon.
\end{aligned}
$$

So the sequence $\{a_j b_j\}$ converges to $\alpha\beta$.

Part **(4)** is proved in a similar fashion and we leave the details as an exercise.   □

POINT OF CONFUSION 2.9 You were probably puzzled by the choice of N and $\widetilde{N}$ in the proof of part **(3)** of Proposition 2.8—where did the number $\epsilon/(2M + 2|\beta|)$ come from? The answer of course becomes obvious when we read on further in the proof. So the lesson here is that a proof is constructed backward: you look to the end of the proof to see what you need to specify earlier on. Skill in these matters can come only with practice.

When discussing the convergence of a sequence, we often find it inconvenient to deal with the definition of convergence as given. For this definition makes reference to the number to which the sequence is supposed to converge, and we often do not know this number in advance. Would it not be useful to be able to decide whether a series converges *without knowing to what limit it converges*?

Definition 2.10 Let $\{a_j\}$ be a sequence of real numbers. We say that the sequence satisfies the *Cauchy criterion* (A. L. Cauchy, 1789-1857)—more briefly, that the sequence is *Cauchy*—if, for each $\epsilon > 0$, there is an integer $N > 0$ such that if $j, k > N$ then $|a_j - a_k| < \epsilon$.

Notice that the concept of a sequence being Cauchy simply makes precise the notion of the elements of the sequence **(i)** *getting* close together and **(ii)** *staying* close together.

Lemma 2.11 *Every Cauchy sequence is bounded.*

Proof: Let $\epsilon = 1 > 0$. There is an integer $N > 0$ such that $|a_j - a_k| < \epsilon = 1$ whenever $j, k > N$. Thus if $j \geq N + 1$ we have

$$
\begin{aligned}
|a_j| &= |a_{N+1} + (a_j - a_{N+1})| \\
&\leq |a_{N+1}| + |a_j - a_{N+1}| \\
&\leq |a_{N+1}| + 1 \equiv K.
\end{aligned}
$$

Let $L = \max\{|a_1|, |a_2|, \ldots, |a_N|\}$. If j is any natural number, then either $1 \leq j \leq N$, in which case $|a_j| \leq L$, or else $j > N$, in which case $|a_j| \leq K$. Set $M = \max\{K, L\}$. Then, for any j, $|a_j| \leq M$ as required.   □

Theorem 2.12 *Let $\{a_j\}$ be a sequence of real numbers. The sequence is Cauchy if and only if it converges to some limit α.*

Proof: First assume that the sequence converges to a limit α. Let $\epsilon > 0$. Choose, by definition of convergence, an integer $N > 0$ such that if $j > N$ then $|a_j - \alpha| < \epsilon/2$. If $j, k > N$ then

$$|a_j - a_k| \leq |a_j - \alpha| + |\alpha - a_k| < \frac{\epsilon}{2} + \frac{\epsilon}{2} = \epsilon.$$

So the sequence is Cauchy.

Conversely, suppose that the sequence is Cauchy. Define

$$S = \{x \in \mathbb{R} : x < a_j \text{ for all but finitely many } j\}.$$

[**Hint:** You might find it helpful to think of this set as

$$S = \{x \in \mathbb{R} : \text{there is a positive integer } k \text{ such that } x < a_j \text{ for all } j \geq k\}.]$$

By the lemma, the sequence $\{a_j\}$ is bounded by some number M. If x is a real number less than $-M$, then $x \in S$, so S is nonempty. Also S is bounded above by M. Let $\alpha = \sup S$. Then α is a well-defined real number, and we claim that α is the limit of the sequence $\{a_j\}$.

To see this, let $\epsilon > 0$. Choose an integer $N > 0$ such that $|a_j - a_k| < \epsilon/2$ whenever $j, k > N$. Notice that this last inequality implies that

$$|a_j - a_{N+1}| < \epsilon/2 \text{ when } j \geq N + 1 \qquad (2.12.1)$$

hence

$$a_j > a_{N+1} - \epsilon/2 \text{ when } j \geq N + 1.$$

Thus $a_{N+1} - \epsilon/2 \in S$ and it follows that

$$\alpha \geq a_{N+1} - \epsilon/2. \qquad (2.12.2)$$

Line (2.12.1) also shows that

$$a_j < a_{N+1} + \epsilon/2 \text{ when } j \geq N+1.$$

Thus $a_{N+1} + \epsilon/2 \notin S$ and

$$\alpha \leq a_{N+1} + \epsilon/2. \tag{2.12.3}$$

Combining lines (2.12.2) and (2.12.3) gives

$$|\alpha - a_{N+1}| \leq \epsilon/2. \tag{2.12.4}$$

But then line (2.12.4) yields, for $j > N$, that

$$|\alpha - a_j| \leq |\alpha - a_{N+1}| + |a_{N+1} - a_j| < \epsilon/2 + \epsilon/2 = \epsilon.$$

This proves that the sequence $\{a_j\}$ converges to α, as claimed.   □

POINT OF CONFUSION 2.13 Any convergent sequence is Cauchy. And, in the real number system, any Cauchy sequence is convergent. So why do we have both concepts? The answer is easy. Convergent sequences are what we are really interested in. But the concept of Cauchy sequence helps us to identify them.

Definition 2.14 Let $\{a_j\}$ be a sequence of real numbers. The sequence is said to be *increasing* if $a_1 \leq a_2 \leq \ldots$. It is *decreasing* if $a_1 \geq a_2 \geq \ldots$.

A sequence is said to be *monotone* if it is either increasing or decreasing.

Proposition 2.15 *If $\{a_j\}$ is an increasing sequence which is bounded above—$a_j \leq M < \infty$ for all j—then $\{a_j\}$ is convergent. If $\{b_j\}$ is a decreasing sequence which is bounded below—$b_j \geq K > -\infty$ for all j—then $\{b_j\}$ is convergent.*

Proof: Let $\epsilon > 0$. Let $\alpha = \sup a_j < \infty$. By definition of supremum, there is an integer N so that $|a_N - \alpha| < \epsilon$. Then, if $\ell \geq N+1$, we have $a_N \leq a_\ell \leq \alpha$ hence $|a_\ell - \alpha| < \epsilon$. Thus the sequence converges to α.

The proof for decreasing sequences is similar and we omit it.   □

Remark 2.16 Let $a_1 = \sqrt{2}$ and set $a_{j+1} = \sqrt{2 + a_j}$ for $j \geq 1$. You can verify that $\{a_j\}$ is increasing and bounded above (by 4 for example). What is its limit (which is guaranteed to exist by the proposition)?   ∎

A proof very similar to that of the proposition gives the following useful fact:

Corollary 2.17 *Let S be a nonempty set of real numbers which is bounded above and below. Let β be its supremum and α its infimum. If $\epsilon > 0$ then there are $s, t \in S$ such that $|s - \beta| < \epsilon$ and $|t - \alpha| < \epsilon$.*

Proof: This is essentially a restatement of the proof of the proposition. $\square$

We conclude the section by recording one of the most useful results for calculating the limit of a sequence:

Proposition 2.18 (The Pinching Principle) *Let $\{a_j\}, \{b_j\}$, and $\{c_j\}$ be sequences of real numbers satisfying*

$$a_j \leq b_j \leq c_j$$

for every j sufficiently large. If

$$\lim_{j \to \infty} a_j = \lim_{j \to \infty} c_j = \alpha$$

for some real number α, then

$$\lim_{j \to \infty} b_j = \alpha.$$

Proof: This proof is requested of you in the exercises. $\square$

POINT OF CONFUSION 2.19 The Pinching Principle gives us a way to compare an unknown sequence with two known sequences. It is a powerful tool. But it again drives home the point that we need to have knowledge of a library of reference sequences.

EXAMPLE 2.20 Define

$$a_j = \frac{\sin j \cos 2j}{j^2}.$$

Then

$$0 \leq |a_j| \leq \frac{1}{j^2}.$$

It is clear that

$$\lim_{j \to \infty} 0 = 0$$

and

$$\lim_{j \to \infty} \frac{1}{j^2} = 0.$$

Therefore

$$\lim_{j \to \infty} |a_j| = 0$$

so that

$$\lim_{j \to \infty} a_j = 0. \qquad \blacksquare$$

A Look Back

1. Speaking intuitively, what does it mean for a sequence to converge?

2. Speaking intuitively, what does it mean for a sequence to be Cauchy?

3. What is a bounded sequence?

4. What is a monotone sequence?

5. Why does any given sequence have at most one limit?

Exercises

1. Prove parts **(2)** and **(4)** of Proposition 2.6.

2. Prove the following result, which we have used without comment in the text: Let S be a set of real numbers which is bounded above and let $t = \sup S$. For any $\epsilon > 0$ there is an element $s \in S$ such that $t - \epsilon < s \le t$. [**Hint:** Notice that this result makes good intuitive sense: the elements of S should become arbitrarily close to the supremum t, otherwise there would be enough room to decrease the value of t and make the supremum even smaller.] Formulate and prove a similar result for the infimum.

3. Prove Proposition 2.18.

4. Let $a_1, a_2 > 0$ and for $j \ge 3$ define $a_j = a_{j-1} + a_{j-2}$. Show that this sequence cannot converge to a finite limit.

5. Suppose a sequence $\{a_j\}$ has the property that, for every natural number N, there is a j_N such that $a_{j_N} = a_{j_N+1} = \cdots = a_{j_N+N}$. In other words, the sequence has arbitrarily long repetitive strings. Does it follow that the sequence converges?

6. Let γ be an irrational real number and let a_j be a sequence of rational numbers converging to γ. Suppose that each a_j is a fraction expressed in lowest terms: $a_j = \alpha_j/\beta_j$. Prove that the β_j tend to ∞.

7. Use the integral of $1/(1 + t^2)$, together with Riemann sums (ideas which you know from calculus, and which we shall treat rigorously later in the book), to develop a scheme for calculating the digits of π.

8. Does the sequence $a_j = (-1)^j j^2/(j^2 + 1)$ converge? If so, to what limit?

9. Does the sequence $a_j = e^j/(e^{2j} + j^2)$ converge? If so, to what limit?

10. Define $a_1 = 1$, $a_2 = 1$, $a_3 = a_1 + a_2$, and $a_j = a_{j-2} + a_{j-1}$ for $j \ge 4$. Does this sequence converge? If so, to what limit?

11. Give an example of a sequence of rational numbers that converges to π. Give an example of a sequence of irrational numbers that converges to 2.

12. Show that, between any two distinct rational numbers, there is an irrational number.

13. Show that, between any two distinct irrational numbers, there is a rational number.

14. Prove that the sum of two irrational numbers can be either rational or irrational. Prove that the product of two irrational numbers can be either rational or irrational.

15. Give an example of a sequence $\{a_j\}$ that converges to $+\infty$ but so that it is *not* the case that $a_1 \le a_2 \le a_3 \le \cdots$.

2.2 Subsequences

Preliminary Remarks

A subsequence of a given sequence is a "junior" sequence that lives inside the parent sequence. Many of the most important ideas in analysis, including closure and compactness, are formulated in terms of subsequences. It is a rather slippery idea, but one that you need to master.

Let $\{a_j\}$ be a given sequence. If

$$0 < j_1 < j_2 < \cdots$$

are positive integers then the function

$$k \mapsto a_{j_k}$$

is called a *subsequence* of the given sequence. We usually write the subsequence as

$$\{a_{j_k}\}_{k=1}^{\infty} \quad \text{or} \quad \{a_{j_k}\}.$$

EXAMPLE 2.21 Consider the sequence

$$\{2^j\} = \{2, 4, 8, \dots\}.$$

Then the sequence

$$\{2^{2k}\} = \{4, 16, 64, \dots\} \tag{2.21.1}$$

is a subsequence. Notice that the subsequence contains a subcollection of elements of the original sequence *in the same order*. In this example, $j_k = 2k$.

Another subsequence is

$$\{2^{(2^k)}\} = \{4, 16, 256, \dots\}.$$

In this instance, it holds that $j_k = 2^k$. Notice that this new subsequence is in fact a subsequence of the first subsequence (2.21.1). That is, it is a sub-subsequence of the original sequence $\{2^j\}$. ∎

Proposition 2.22 *If $\{a_j\}$ is a convergent sequence with limit α, then every subsequence converges to the limit α.*

Conversely, if a sequence $\{b_j\}$ has the property that each of its subsequences is convergent then $\{b_j\}$ itself is convergent.

Proof: Assume $\{a_j\}$ is convergent to a limit α, and let $\{a_{j_k}\}$ be a subsequence. Let $\epsilon > 0$ and choose $N > 0$ such that $|a_j - \alpha| < \epsilon$ whenever $j > N$. Now if $k > N$ then $j_k > N$ hence $|a_{j_k} - \alpha| < \epsilon$. Therefore, by definition, the subsequence $\{a_{j_k}\}$ also converges to α.

The converse is trivial, simply because the sequence is a subsequence of itself. $\square$

POINT OF CONFUSION 2.23 A subsequence of a given sequence is also a *subset* of that sequence. But it is much more than that, because it maintains the same order of terms.

Now we present one of the most fundamental theorems of basic real analysis (due to B. Bolzano, 1781-1848, and K. Weierstrass, 1815-1897).

Theorem 2.24 (Bolzano-Weierstrass) *Let $\{a_j\}$ be a bounded sequence in $\mathbb{R}$. Then there is a subsequence which converges.*

Proof: Say that $|a_j| \leq M$ for every j. We may assume that $M > 0$.

One of the two intervals $[-M, 0]$ and $[0, M]$ must contain infinitely many elements of the sequence. Say that $[0, M]$ does. Choose a_{j_1} to be one of the infinitely many sequence elements in $[0, M]$.

Next, one of the intervals $[0, M/2]$ and $[M/2, M]$ must contain infinitely many elements of the sequence. Say that it is $[0, M/2]$. Choose an element $a_{j_2} \in [0, M/2]$ with $j_2 > j_1$. Continue in this fashion, halving the interval, choosing a half with infinitely many sequence elements, and selecting the next subsequential element from that half.

Let us analyze the resulting subsequence. Notice that $|a_{j_1} - a_{j_2}| \leq M$ since both elements belong to the interval $[0, M]$. Likewise, $|a_{j_2} - a_{j_3}| \leq M/2$ since both elements belong to $[0, M/2]$. In general, $|a_{j_k} - a_{j_{k+1}}| \leq 2^{-k+1} \cdot M$ for each $k \in \mathbb{N}$. Now let $\epsilon > 0$. Choose an integer $N > 0$ such that $2^{-N} < \epsilon/(4M)$.

Then, for any $m > l > N$ we have

$$
\begin{aligned}
|a_{j_l} - a_{j_m}| &= |(a_{j_l} - a_{j_{l+1}}) + (a_{j_{l+1}} - a_{j_{l+2}}) + \cdots + (a_{j_{m-1}} - a_{j_m})| \\
&\leq |a_{j_l} - a_{j_{l+1}}| + |a_{j_{l+1}} - a_{j_{l+2}}| + \cdots + |a_{j_{m-1}} - a_{j_m}| \\
&\leq 2^{-l+1} \cdot M + 2^{-l} \cdot M + \cdots + 2^{-m+2} \cdot M \\
&= \left(2^{-l+1} + 2^{-l} + 2^{-l-1} + \cdots + 2^{-m+2}\right) \cdot M \\
&= \big((2^{-l+2} - 2^{-l+1}) + (2^{-l+1} - 2^{-l}) + \cdots \\
&\qquad + (2^{-m+3} - 2^{-m+2})\big) \cdot M \\
&= \left(2^{-l+2} - 2^{-m+2}\right) \cdot M \\
&< 2^{-l+2} \cdot M \\
&< 4 \cdot \frac{\epsilon}{4M} \cdot M \\
&= \epsilon.
\end{aligned}
$$

We see that the subsequence $\{a_{j_k}\}$ is Cauchy, so it converges. $\qquad\square$

POINT OF CONFUSION 2.25 Of course it is not true that every bounded sequence converges. But the Bolzano-Weierstrass theorem is a good substitute result. Often, in practice, all that we need is a convergent subsequence.

Remark 2.26 The Bolzano-Weierstrass theorem is a generalization of our result from the last section about increasing sequences which are bounded above (resp. decreasing sequences which are bounded below). For such a sequence is surely bounded above *and* below (why?). So it has a convergent subsequence. And thus it follows easily that the entire sequence converges. Details are left as an exercise.

It is a fact—which you can verify for yourself—that *any* real sequence has a monotone subsequence. This fact implies Bolzano-Weierstrass.

POINT OF CONFUSION 2.27 In this text we have not yet given a rigorous definition of the function $\sin x$ (see Section 9.3). However, just for the moment, use the definition you learned in calculus class and consider the sequence $\{\sin j\}_{j=1}^{\infty}$. Notice that the sequence is bounded in absolute value by 1. The Bolzano-Weierstrass theorem guarantees that there is a convergent subsequence, even though it would be very difficult to say precisely what that convergent subsequence is. $\qquad\square$

A Look Back

1. What is a subsequence?

2. How can it happen that a sequence does not converge but one of its subsequences does converge?

3. Is it possible for two different subsequences to converge to two different limits?

4. Is it possible for two different subsequences to converge to the same limit?

Exercises

1. Use the Bolzano-Weierstrass theorem to show that every increasing sequence that is bounded above converges.

2. Give an example of a sequence of rational numbers with the property that, for any real number α, there is a subsequence converging to α.

3. Let $x_1 = 2$. For $j \geq 1$, set

$$x_{j+1} = x_j - \frac{x_j^2 - 2}{2x_j}.$$

Show that the sequence $\{x_j\}$ is decreasing and bounded below. What is its limit?

4. The sequence $\{\cos j + \sin j\}$ has a convergent subsequence. Explain why. Can you say what that subsequence is?

5. Give an example of a sequence that is bounded below but does not have a convergent subsequence. Give an example of a sequence that is bounded above but does not have a convergent subsequence.

*   6. Provide the details of the assertion, made in the text, that the sequence $\{\cos j\}$ is dense in the interval $[-1, 1]$. [**Hint:** See Exercise **14** in Section 1.1.]

*   7. Give another proof of the Bolzano-Weierstrass theorem as follows. If $\{a_j\}$ is a bounded sequence let $b_j = \inf\{a_j, a_{j+1}, \dots\}$. Then each b_j is finite, $b_1 \leq b_2 \leq \dots$, and $\{b_j\}$ is bounded above. Now use Proposition 2.15.

*   8. Let $S = \{0, 1, 1/2, 1/3, 1/4, \dots\}$. Give an example of a sequence $\{a_j\}$ with the property that, for each $s \in S$, there is a subsequence converging to s, but no subsequence converges to any limit not in S.

*   9. Give an example of a sequence with infinitely many distinct subsequences that converge to π.

10. Give an example of a sequence which does not converge, but which has infinitely many different subsequences that do converge.

11. Prove that a sequence $\{a_j\}$ converges if and only if every subsequence has a subsequence that converges.

2.3 Lim sup and Lim inf

Preliminary Remarks

While our interest in sequences is in their limits, it is a fact that most sequences do not have a limit. So what can we do? Is there a substitute idea?

In fact there is, and that is the idea of lim sup (limit superior) and lim inf (limit inferior). These are, in effect, the greatest limit of any subsequence and the least limit of any subsequence.

You can see that this new set of ideas combines many of the earlier ideas. But it leads to greater depth and insight, and it is important.

Convergent sequences are useful objects, but the unfortunate truth is that most sequences do not converge. Nevertheless, we would like to have a language for discussing the asymptotic behavior of *any* real sequence $\{a_j\}$ as $j \to \infty$. That is the purpose of the concepts of "limit superior" (or "upper limit") and "limit inferior" (or "lower limit"). It should be stressed that *any sequence whatever* has a lim sup and a lim inf.

Definition 2.28 Let $\{a_j\}$ be a sequence of real numbers. For each j let

$$A_j = \inf\{a_j, a_{j+1}, a_{j+2}, \dots\}.$$

Then $\{A_j\}$ is an increasing sequence (since, as j increases, we are taking the infimum of a smaller and smaller set of numbers), so it has a limit. We define the *limit infimum* of $\{a_j\}$ to be

$$\liminf a_j = \lim_{j\to\infty} A_j.$$

It is common to refer to this number as the lim inf of the sequence.

Likewise, let

$$B_j = \sup\{a_j, a_{j+1}, a_{j+2}, \dots\}.$$

Then $\{B_j\}$ is a decreasing sequence (since, as j increases, we are taking the supremum of a smaller and smaller set of numbers), so it has a limit. We define the *limit supremum* of $\{a_j\}$ to be

$$\limsup a_j = \lim_{j\to\infty} B_j.$$

It is common to refer to this number as the lim sup of the sequence.

POINT OF CONFUSION 2.29 Notice that the lim sup or lim inf of a sequence can be $\pm\infty$. For instance, the sequence $a_j = j^2 - j$ has lim sup equal to $+\infty$. The sequence $-2j + 6$ has lim inf equal to $-\infty$.

Remark 2.30 What is the intuitive content of this definition? For each j, A_j picks out the greatest lower bound of the sequence in the jth position or later. So the sequence $\{A_j\}$ should tend to the *smallest* possible limit of any subsequence of $\{a_j\}$.

Likewise, for each j, B_j picks out the least upper bound of the sequence in the jth position or later. So the sequence $\{B_j\}$ should tend to the *greatest* possible limit of any subsequence of $\{a_j\}$. We shall make these remarks more precise in Proposition 2.33 below.

Notice that it is implicit in the definition that *every* real sequence has a limit supremum and a limit infimum.

POINT OF CONFUSION 2.31 It is important to keep in mind that the lim sup of a given sequence is in fact associated to a subsequence. And so is the lim inf. In practice we think of the lim sup as the limit of the "greatest" subsequence and the lim inf as the limit of the "least" subsequence.

A further comment is that we can talk about the limit infimum of a sequence even when the sequence is *not* bounded below. But then some or all of the A_j may be $-\infty$ and the limit infimum may be $-\infty$. Likewise we may discuss the limit supremum of a sequence that is not bounded above.

EXAMPLE 2.32 Consider the sequence $\{(-1)^j\}$. Of course this sequence does not converge. Let us calculate its lim sup and lim inf.

Referring to the definition, we have that $A_j = -1$ for every j. So

$$\liminf (-1)^j = \lim (-1) = -1.$$

Similarly, $B_j = +1$ for every j. Therefore

$$\limsup (-1)^j = \lim (+1) = +1.$$

As we predicted in the remark, the lim inf is the least subsequential limit, and the lim sup is the greatest subsequential limit.   ∎

Now let us prove the characterizing property of lim sup and lim inf to which we have been alluding.

Proposition 2.33 *Let $\{a_j\}$ be a sequence of real numbers. Let $\beta = \limsup_{j\to\infty} a_j$ and $\alpha = \liminf_{j\to\infty} a_j$. If $\{a_{j_\ell}\}$ is any subsequence of the given sequence then*

$$\alpha \leq \liminf_{\ell\to\infty} a_{j_\ell} \leq \limsup_{\ell\to\infty} a_{j_\ell} \leq \beta.$$

Moreover, there is a subsequence $\{a_{j_k}\}$ such that

$$\lim_{k\to\infty} a_{j_k} = \alpha$$

and another sequence $\{a_{n_k}\}$ such that

$$\lim_{k\to\infty} a_{n_k} = \beta.$$

Proof: For simplicity in this proof we assume that the lim sup and lim inf are finite.

We begin by considering the lim inf. There is a $j_1 \geq 1$ such that $|A_1 - a_{j_1}| < 2^{-1}$. We choose j_1 to be as small as possible. Next, we choose j_2, necessarily greater than j_1, such that j_2 is as small as possible and $|a_{j_2} - A_2| < 2^{-2}$. Continuing in this fashion, we select $j_k > j_{k-1}$ such that $|a_{j_k} - A_k| < 2^{-k}$, etc.

Recall that $A_k \to \alpha = \liminf_{j\to\infty} a_j$. Now fix $\epsilon > 0$. If N is an integer so large that $k > N$ implies that $|A_k - \alpha| < \epsilon/2$ and also that $2^{-N} < \epsilon/2$ then, for such k, we have

$$
\begin{aligned}
|a_{j_k} - \alpha| &\leq |a_{j_k} - A_k| + |A_k - \alpha| \\
&< 2^{-k} + \frac{\epsilon}{2} \\
&< \frac{\epsilon}{2} + \frac{\epsilon}{2} \\
&= \epsilon.
\end{aligned}
$$

Thus the subsequence $\{a_{j_k}\}$ converges to α, the lim inf of the given sequence. A similar construction gives a (different) subsequence $\{a_{n_k}\}$ converging to β, the lim sup of the given sequence.

Now let $\{a_{j_\ell}\}$ be *any* subsequence of the sequence $\{a_j\}$. Let β^* be the lim sup of this subsequence. Then, by the first part of the proof, there is a subsequence $\{a_{j_{\ell_m}}\}$ such that

$$\lim_{m\to\infty} a_{j_{\ell_m}} = \beta^*.$$

But $a_{j_{\ell_m}} \leq B_{j_{\ell_m}}$ by the very definition of the Bs. Thus

$$\beta^* = \lim_{m\to\infty} a_{j_{\ell_m}} \leq \lim_{m\to\infty} B_{j_{\ell_m}} = \beta$$

or

$$\limsup_{\ell\to\infty} a_{j_\ell} \leq \beta,$$

as claimed. A similar argument shows that

$$\liminf_{l\to\infty} a_{j_l} \geq \alpha.$$

This completes the proof of the proposition. $\square$

Corollary 2.34 *If $\{a_j\}$ is a sequence and $\{a_{j_k}\}$ is a convergent subsequence then*

$$\liminf_{j\to\infty} a_j \le \lim_{k\to\infty} a_{j_k} \le \limsup_{j\to\infty} a_j.$$

We close this section with a fact that is analogous to one for the supremum and infimum. Its proof is analogous to arguments we have seen before.

Proposition 2.35 *Let $\{a_j\}$ be a sequence and set $\limsup a_j = \beta$ and $\liminf a_j = \alpha$. Assume that α, β are finite real numbers. Let $\epsilon > 0$. Then there are arbitrarily large j such that $a_j > \beta - \epsilon$. Also there are arbitrarily large k such that $a_k < \alpha + \epsilon$.*

POINT OF CONFUSION 2.36 Given a sequence, there are always elements of that sequence which are arbitrarily close to the lim sup, and there are elements of the sequence which are arbitrarily close to the lim inf. It is often this particular property of the concepts which is most useful.

A Look Back

1. Intuitively speaking, what is a lim sup?
2. Intuitively speaking, what is a lim inf?
3. In what sense is the lim sup of a sequence greater than or equal to the lim inf of that sequence?
4. What does it mean when the lim sup and lim inf are equal?

Exercises

1. Consider $\{a_j\}$ both as a sequence and as a set. How are the lim sup and the sup related? How are the lim inf and the inf related? Give examples.

2. Prove the last proposition in this section.

3. Let $\{a_j\}$ be a sequence of positive numbers with positive liminf. Then prove that
$$\limsup 1/a_j = 1/\liminf a_j.$$

4. What are the limsup and liminf of $|\sin j|^{\sin j}$?

5. Prove that if $\limsup a_j = \liminf a_j$ then the sequence $\{a_j\}$ converges.

6. Let $a < b$ be real numbers. Give an example of a real sequence whose lim sup is b and whose lim inf is a.

7. How is $\limsup(a_j + b_j)$ related to $\limsup a_j$ and $\limsup b_j$?

8. Explain why we do not consider lim sup and lim inf for complex numbers.

9. How are the limsup and liminf of $a_j \cdot b_j$ related to the limsup and liminf of a_j and b_j?

10. How are the lim sup and lim inf of $\{a_j\}$ related to the lim sup and lim inf of $\{-a_j\}$?

11. What is $\limsup_{j\to\infty} \cos j$? What is $\limsup_{j\to\infty} \sin j$?

2.4 Some Special Sequences

Preliminary Remarks

One of the ways that we understand a new sequence is to compare it with a known sequence. Thus we need a library of "known" sequences that we can use as the basis for our studies. This section begins to assemble such a library.

We often obtain information about a new sequence by comparison with a sequence that we already know. Thus it is well to have a catalogue of fundamental sequences which provide a basis for comparison.

EXAMPLE 2.37 Fix a real number a. The sequence $\{a^j\}$ is called a *power sequence*. If $-1 < a < 1$ then the sequence converges to 0. If $a = 1$ then the sequence is a constant sequence and converges to 1. If $a > 1$ then the sequence diverges to $+\infty$. Finally, if $a \leq -1$, then the sequence diverges. ∎

Recall that, in Section 1.1, we discussed the existence of nth roots of positive real numbers. If $\alpha > 0, m \in \mathbb{Z}$, and $n \in \mathbb{N}$, then we may define

$$\alpha^{m/n} = (\alpha^m)^{1/n} .$$

Thus we may talk about rational powers of a positive number. Next, if $\beta \in \mathbb{R}$, then we may define

$$\alpha^\beta = \sup\{\alpha^q : q \in \mathbb{Q}, q < \beta\} .$$

Thus we can define *any real power* of a positive real number. Some of the exercises ask you to verify several basic properties of these exponentials.

Lemma 2.38 *If $\alpha > 1$ is a real number and $\beta > 0$ then $\alpha^\beta > 1$.*

Proof: Let q be a positive rational number which is less than β. Say that $q = m/n$, with m, n integers. It is obvious that $\alpha^m > 1$ and hence that $(\alpha^m)^{1/n} > 1$. Since α^β majorizes this last quantity, we are done. □

EXAMPLE 2.39 Fix a real number α and consider the sequence $\{j^\alpha\}$. If $\alpha > 0$ then it is easy to see that $j^\alpha \to +\infty$: to verify this assertion fix $M > 0$ and take the number N to be the first integer after $M^{1/\alpha}$.

If $\alpha = 0$ then j^α is a constant sequence, identically equal to 1.

If $\alpha < 0$ then $j^\alpha = 1/j^{-\alpha}$. The denominator of this last expression tends to $+\infty$ hence the sequence j^α tends to 0. ∎

EXAMPLE 2.40 The sequence $\{j^{1/j}\}$ converges to 1. In fact, consider the expressions $\alpha_j = j^{1/j} - 1 > 0$. We have that

$$j = (\alpha_j + 1)^j \geq \frac{j(j-1)}{2}(\alpha_j)^2$$

(the latter being just one term from the binomial expansion). Thus

$$0 < \alpha_j \leq \sqrt{2/(j-1)}$$

as long as $j \geq 2$. It follows that $\alpha_j \to 0$ or $j^{1/j} \to 1$. ∎

EXAMPLE 2.41 Let α be a positive real number. Then the sequence $\alpha^{1/j}$ converges to 1. To see this, first note that the case $\alpha = 1$ is trivial, and the case $\alpha > 1$ implies the case $\alpha < 1$ (by taking reciprocals). So we concentrate on $\alpha > 1$. But then we have

$$1 < \alpha^{1/j} < j^{1/j}$$

when $j > \alpha$. Since $j^{1/j}$ tends to 1, Proposition 2.18 applies and the proof is complete. ∎

EXAMPLE 2.42 Let $\lambda > 1$ and let α be real. Then the sequence

$$\left\{ \frac{j^\alpha}{\lambda^j} \right\}_{j=1}^\infty$$

converges to 0.

To see this, fix an integer $k > \alpha$ and consider $j > 2k$. [Notice that k is fixed once and for all but j will be allowed to tend to $+\infty$ at the appropriate moment.] Writing $\lambda = 1 + \mu, \mu > 0$, we have that

$$\lambda^j = (1 + \mu)^j > \frac{j(j-1)(j-2)\cdots(j-k+1)}{k(k-1)(k-2)\cdots 2 \cdot 1}\mu^k \cdot 1^{j-k}.$$

Of course this comes from picking out the kth term of the binomial expansion for $(1 + \mu)^j$. Notice that, since $j > 2k$, then each of the expressions $j, (j - 1), \ldots (j - k + 1)$ in the numerator on the right exceeds $j/2$. Thus

$$\lambda^j > \frac{j^k}{2^k \cdot k!} \cdot \mu^k$$

and

$$0 < \frac{j^\alpha}{\lambda^j} < j^\alpha \cdot \frac{2^k \cdot k!}{j^k \cdot \mu^k} = \frac{j^{\alpha-k} \cdot 2^k \cdot k!}{\mu^k}.$$

Since $\alpha - k < 0$, the right side tends to 0 as $j \to \infty$. ∎

POINT OF CONFUSION 2.43 We should always bear in mind that expressions of the form α^j grow *much faster* than expressions of the form j^β (when $\alpha > 1$, $\beta > 1$, for instance). That's because exponentials grow faster than polynomials. Keeping this fact in focus helps in evaluating the limits of particular sequences.

EXAMPLE 2.44 The sequence

$$\left\{ \left(1 + \frac{1}{j}\right)^j \right\}$$

converges. In fact it is increasing and bounded above. Use the Binomial Expansion to prove this assertion. The limit of the sequence is the number that we shall later call e (in honor of Leonhard Euler, 1707-1783, who first studied it in detail). We shall study this sequence in detail later in the book.
∎

EXAMPLE 2.45 The sequence

$$\left(1 - \frac{1}{j}\right)^j$$

converges to $1/e$, where the definition of e is given in the last example. More generally, the sequence

$$\left(1 + \frac{x}{j}\right)^j$$

converges to e^x (here e^x is defined as in the discussion following Example 2.37 above).
∎

A Look Back

1. What is a sequence that converges to Euler's number e?

2. What is a sequence that converges to π?

3. Is there a power sequence that converges to 2?

4. Is there a sequence of rational numbers that converges to Euler's number e?

Exercises

1. Let α be a positive real number and let $p/q = m/n$ be two different representations of the same rational number r. Prove that

$$(\alpha^m)^{1/n} = (\alpha^p)^{1/q} .$$

Also prove that
$$(\alpha^{1/n})^m = (\alpha^m)^{1/n}.$$

If β is another positive real and γ is any real then prove that
$$(\alpha \cdot \beta)^\gamma = \alpha^\gamma \cdot \beta^\gamma.$$

2. Discuss the convergence of the sequence $\{(1/j)^{1/j}\}_{j=1}^{\infty}$.

3. Discuss the convergence of the sequence $(1 + x/j)^j$.

4. Prove that the exponential, as defined in this section, satisfies
$$(a^b)^c = a^{bc} \qquad \text{and} \qquad a^b a^c = a^{b+c}.$$

5. Discuss the convergence of the sequence $j^j/(2j)!$.

6. For which values of $\alpha > 0$ and $\beta > 0$ does the sequence
$$a_j = \frac{\alpha^j}{j^\beta}$$

converge? What about the sequence
$$b_j = \frac{j^\beta}{\alpha^j} \,?$$

7. Give a formula for the Fibinacci sequence.

8. Discuss convergence of the sequence $j^{1/j}$.

9. Discuss convergence of the sequence $(1 + 1/j^2)^j$.

* * 10. Consider the sequence given by
$$a_j = \left[1 + \frac{1}{2} + \frac{1}{3} + \cdots + \frac{1}{j} \right] - \log j.$$

Use a picture (remember that log is the antiderivative of $1/x$) to give a convincing argument that the sequence $\{a_j\}$ converges. The limit number is called γ. This number was first studied by Euler. It arises in many different contexts in analysis and number theory.

As a challenge problem, show that
$$|a_j - \gamma| \le \frac{C}{j}$$

for some universal constant $C > 0$.

It is not known whether γ is rational or irrational.

* * 11. A sequence is defined by the rule $a_0 = 2$, $a_1 = 1$, and $a_j = 3a_{j-1} - a_{j-2}$. Find a formula for a_j.

Chapter 3

Series of Numbers

3.1 Convergence of Series

Preliminary Remarks

A series is an infinite sum. We think of the series as the limit of the sequence of its partial sums. While at first a bit confusing, this approach avoids many conundrums and redundancies that have plagued the history of series.

Of course we care about whether a series converges, and what it converges to. This chapter will be devoted to the study of these questions.

In this section we will use standard summation notation:

$$\sum_{j=m}^{n} a_j \equiv a_m + a_{m+1} + \cdots + a_n .$$

EXAMPLE 3.1 We calculate two sample sums:

$$\sum_{j=2}^{5} j^2 = 2^2 + 3^2 + 4^2 + 5^2 = 54 ,$$

$$\sum_{j=4}^{8} 2^{-j} = 2^{-4} + 2^{-5} + 2^{-6} + 2^{-7} + 2^{-8} = \frac{2^5 - 1}{2^8} . \qquad \blacksquare$$

A series is an infinite sum. One of the most effective ways to handle an infinite process in mathematics is with a limit. This consideration leads to the following definition:

Definition 3.2 The formal expression

$$\sum_{j=1}^{\infty} a_j \, ,$$

where the a_j are real numbers, is called a *series*. For $N = 1, 2, 3, \ldots$, the expression

$$S_N = \sum_{j=1}^{N} a_j = a_1 + a_2 + \ldots a_N$$

is called the Nth *partial sum* of the series. In case

$$\lim_{N \to \infty} S_N$$

exists and is finite we say that the series *converges*. The limit of the partial sums is called the *sum* of the series. If the series does not converge, then we say that the series *diverges*.

POINT OF CONFUSION 3.3 Notice that the question of convergence of a series, which should be thought of as an *addition process*, reduces to a question about the *sequence* of partial sums.

An obvious way to be misled here is to confuse the roles of sequences and series. The way that we analyze a series is that we think of it as a sequence of partial sums. The series converges if and only if the sequence of partial sums converges.

EXAMPLE 3.4 Consider the series

$$\sum_{j=1}^{\infty} 2^{-j} \, .$$

The Nth partial sum for this series is

$$S_N = 2^{-1} + 2^{-2} + \cdots + 2^{-N} \, .$$

In order to determine whether the sequence $\{S_N\}$ has a limit, we rewrite S_N as

$$\begin{aligned} S_N &= \left(2^{-0} - 2^{-1}\right) + \left(2^{-1} - 2^{-2}\right) + \cdots \\ &\quad \left(2^{-N+1} - 2^{-N}\right) \, . \end{aligned}$$

The expression on the right of the last equation telescopes (i.e., successive pairs of terms cancel) and we find that

$$S_N = 2^{-0} - 2^{-N} \, .$$

Thus

$$\lim_{N \to \infty} S_N = 2^{-0} = 1 \,.$$

We conclude that the series converges.   ∎

EXAMPLE 3.5 Let us examine the series

$$\sum_{j=1}^{\infty} \frac{1}{j}$$

for convergence or divergence. (This series is commonly called the *harmonic series* because it describes the harmonics in music.) Now

$$S_1 \;\; = \;\; 1 = \frac{2}{2}$$

$$S_2 \;\; = \;\; 1 + \frac{1}{2} = \frac{3}{2}$$

$$S_4 \;\; = \;\; 1 + \frac{1}{2} + \left(\frac{1}{3} + \frac{1}{4} \right)$$

$$\;\; \geq \;\; 1 + \frac{1}{2} + \left(\frac{1}{4} + \frac{1}{4} \right) \geq 1 + \frac{1}{2} + \frac{1}{2} = \frac{4}{2}$$

$$S_8 \;\; = \;\; 1 + \frac{1}{2} + \left(\frac{1}{3} + \frac{1}{4} \right) + \left(\frac{1}{5} + \frac{1}{6} + \frac{1}{7} + \frac{1}{8} \right)$$

$$\;\; \geq \;\; 1 + \frac{1}{2} + \left(\frac{1}{4} + \frac{1}{4} \right) + \left(\frac{1}{8} + \frac{1}{8} + \frac{1}{8} + \frac{1}{8} \right)$$

$$\;\; = \;\; \frac{5}{2} \,.$$

In general this argument shows that

$$S_{2^k} \geq \frac{k+2}{2} \,.$$

The sequence $\{S_N\}$ is increasing since the series contains only positive summands. The fact that the partial sums $S_1, S_2, S_4, S_8, \ldots$ increases without bound shows that the entire sequence of partial sums must increase without bound. We conclude that the series diverges.   ∎

POINT OF CONFUSION 3.6 The harmonic series diverges, but it diverges *very* slowly. For example, the sum of the first million terms of the harmonic series is less than 13.82.

Just as with sequences, we have a Cauchy criterion for series:

Proposition 3.7 *The series $\sum_{j=1}^{\infty} a_j$ converges if and only if, for every $\epsilon > 0$, there is an integer $N > 0$ such that, if $n \geq m > N$, then*

$$\left| \sum_{j=m}^{n} a_j \right| < \epsilon. \qquad (3.7.1)$$

The condition (3.7.1) is called the Cauchy criterion for series.

Proof: Suppose that the Cauchy criterion holds. Pick $\epsilon > 0$ and choose N so large that (3.7.1) holds. If $n \geq m > N$, then

$$|S_n - S_m| = \left| \sum_{j=m+1}^{n} a_j \right| < \epsilon$$

by hypothesis. Thus the sequence $\{S_N\}$ is Cauchy in the sense discussed for sequences in Section 2.1. We conclude that the sequence $\{S_N\}$ converges; by definition, therefore, the series converges.

Conversely, if the series converges then, by definition, the sequence $\{S_N\}$ of partial sums converges. In particular, the sequence $\{S_N\}$ must be Cauchy. Thus, for any $\epsilon > 0$, there is a number $N > 0$ such that if $n \geq m > N$ then

$$|S_n - S_m| < \epsilon.$$

This just says that

$$\left| \sum_{j=m+1}^{n} a_j \right| < \epsilon,$$

and this last inequality is the Cauchy criterion for series.   □

POINT OF CONFUSION 3.8 The Cauchy criterion for series simply says that

$$|S_n - S_m| < \epsilon.$$

So the partial sums are getting closer and closer together.

EXAMPLE 3.9 Let us use the Cauchy criterion to verify that the series

$$\sum_{j=1}^{\infty} \frac{1}{j \cdot (j+1)}$$

converges.

Notice that, if $n \geq m > 1$, then

$$\left| \sum_{j=m}^{n} \frac{1}{j \cdot (j+1)} \right| = \left(\frac{1}{m} - \frac{1}{m+1} \right) + \left(\frac{1}{m+1} - \frac{1}{m+2} \right) + \ldots$$
$$+ \left(\frac{1}{n} - \frac{1}{n+1} \right).$$

The sum on the right plainly telescopes and we have

$$\left| \sum_{j=m}^{n} \frac{1}{j \cdot (j+1)} \right| = \frac{1}{m} - \frac{1}{n+1}.$$

Let $\epsilon > 0$. Let us choose N to be the next integer after $1/\epsilon$. Then, for $n \geq m > N$, we may conclude that

$$\left| \sum_{j=m}^{n} \frac{1}{j \cdot (j+1)} \right| = \frac{1}{m} - \frac{1}{n+1} < \frac{1}{m} < \frac{1}{N} < \epsilon.$$

This is the desired conclusion. ∎

The next result gives a necessary condition for a series to converge. It is a useful device for detecting divergent series, although it can never tell us that a series converges.

Proposition 3.10 (The Zero Test) *If the series*

$$\sum_{j=1}^{\infty} a_j$$

converges then the terms a_j tend to zero as $j \to \infty$.

Proof: Since we are assuming that the series converges, then it must satisfy the Cauchy criterion for series. Let $\epsilon > 0$. Then there is an integer $N > 0$ such that, if $n \geq m > N$, then

$$\left| \sum_{j=m}^{n} a_j \right| < \epsilon. \tag{3.10.1}$$

We take $n = m$ and $m > N$. Then (3.10.1) becomes

$$|a_m| < \epsilon.$$

But this is precisely the conclusion that we desire. □

EXAMPLE 3.11 The series $\sum_{j=1}^{\infty}(-1)^j$ must diverge, *even though its terms appear to be cancelling each other out*. The reason is that the summands do not tend to zero; hence the preceding proposition applies.

Write out several partial sums of this series to see more explicitly that the partial sums are $-1, +1, -1, +1, \ldots$ and hence that the series diverges. ∎

POINT OF CONFUSION 3.12 Series with nonnegative summands are generally much easier to understand, and to analyze, than series with both positive and negative summands. This is because series of the first type converge because of *size of the terms* alone. But series of the second type can and do converge because of cancellation. Cancellation is subtle.

We conclude this section with a necessary and sufficient condition for convergence of a series of nonnegative terms. As with some of our other results on series, it amounts to little more than a restatement of a result on sequences.

Proposition 3.13 *A series*

$$\sum_{j=1}^{\infty} a_j$$

with all $a_j \geq 0$ is convergent if and only if the sequence of partial sums is bounded above.

Proof: Notice that, because the summands are nonnegative, we have

$$S_1 = a_1 \leq a_1 + a_2 = S_2\,,$$

$$S_2 = a_1 + a_2 \leq a_1 + a_2 + a_3 = S_3\,,$$

and in general

$$S_N \leq S_N + a_{N+1} = S_{N+1}\,.$$

Thus the sequence $\{S_N\}$ of partial sums forms an increasing sequence. We know that such a sequence is convergent to a finite limit if and only if it is bounded above (see Section 2.1). This completes the proof.   □

EXAMPLE 3.14 The series $\sum_{j=1}^{\infty} 1$ is divergent since the summands are nonnegative and the sequence of partial sums $\{S_N\} = \{N\}$ is unbounded.

Referring back to Example 3.5, we see that the series $\sum_{j=1}^{\infty} \frac{1}{j}$ diverges because its partial sums are unbounded.

We see from the first example that the series $\sum_{j=1}^{\infty} 2^{-j}$ converges because its partial sums are all bounded above by 1.   ∎

It is frequently convenient to begin a series with summation at $j = 0$ or some other term instead of $j = 1$. All of our convergence results still apply to such a series because of the Cauchy criterion. In other words, the convergence or divergence of a series will depend only on the behavior of its "tail."

A Look Back

1. Intuitively speaking, what does it mean for a series to converge?
2. Describe in words what the Cauchy criterion for series says.
3. Give an example of a convergent series.
4. Give an example of a divergent series.

Exercises

1. Discuss convergence or divergence for each of the following series:

   (a) $\displaystyle\sum_{j=1}^{\infty} \frac{(2^j)^2}{j!}$

   (b) $\displaystyle\sum_{j=1}^{\infty} \frac{(2j)!}{(3j)!}$

   (c) $\displaystyle\sum_{j=1}^{\infty} \frac{j!}{j^j}$

   (d) $\displaystyle\sum_{j=1}^{\infty} \frac{(-1)^j}{3j^2 - 5j + 6}$

   (e) $\displaystyle\sum_{j=1}^{\infty} \frac{2j-1}{3j^2 - 2}$

   (f) $\displaystyle\sum_{j=1}^{\infty} \frac{2j-1}{3j^3 - 2}$

   (g) $\displaystyle\sum_{j=1}^{\infty} \frac{\log(j+1)}{[1 + \log j]^j}$

   (h) $\displaystyle\sum_{j=12}^{\infty} \frac{1}{j \log^3 j}$

   (i) $\displaystyle\sum_{j=2}^{\infty} \frac{\log(2)}{\log j}$

   (j) $\displaystyle\sum_{j=2}^{\infty} \frac{1}{j \log^{1.1} j}$

2. If $b_j > 0$ for every j and if $\sum_{j=1}^{\infty} b_j$ converges then prove that $\sum_{j=1}^{\infty} (b_j)^2$ converges. Prove that the assertion is false if the positivity hypothesis is omitted. How about third powers?

3. If $b_j > 0$ for every j and if $\sum_{j=1}^{\infty} b_j$ converges then prove that $\sum_{j=1}^{\infty} \frac{1}{1+b_j}$ diverges.

4. Let $\sum_{j=1}^{\infty} a_j$ be a divergent series of positive terms. Prove that there exist numbers $b_j, 0 < b_j < a_j$, such that $\sum_{j=1}^{\infty} b_j$ diverges.

   Similarly, let $\sum_{j=1}^{\infty} c_j$ be a convergent series of positive terms. Prove that there exist numbers $d_j, 0 < c_j < d_j$, such that $\sum_{j=1}^{\infty} d_j$ converges.

   Thus we see that there is no "smallest" divergent series and no "largest" convergent series.

5. TRUE or FALSE: If $a_j > c > 0$ and $\sum 1/a_j$ converges, then $\sum a_j$ converges.

6. If $b_j > 0$ and $\sum_j b_j$ converges, then what can you say about $\sum_j b_j/(1 + b_j)$?

7. If $b_j > 0$ and $\sum_j b_j$ diverges, then what can you say about $\sum_j 2^{-j} b_j$?

8. If $b_j > 0$ and $\sum_j b_j$ converges, then what can you say about $\sum_j b_j/j^2$?

9. If $a_j > 0$ and $\sum_j a_j^2$ converges, then what can you say about $\sum_j a_j^4$? How about $\sum_j a_j^3$?

10. Let $b_j > 0$. If $\sum b_j$ converges, then what can you say about convergence of $\sum \sqrt{b_j}$?

11. If $a_j > 0$, $b_j > 0$ and $\sum a_j$ converges and $\sum b_j$ converges, then what can you say about $\sum a_j b_j$?

* 12. Discuss convergence and divergence for the series $\sum_j (\sin j)/j$ and $\sum_j (\sin j)^2/j$.

3.2   Elementary Convergence Tests

┌───┐

Preliminary Remarks

One of the elegant features of the theory of numerical series is that there are several convergence tests that are easy to apply to get specific, concrete information about convergence. While these tests are not exhaustive nor comprehensive, they do give a good deal of useful information.

   The tests that we study in this section are for series of positive terms. More advanced tests will be treated in later sections.

└───┘

As previously noted, a series may converge because its terms diminish in size fairly rapidly (thus causing its partial sums to grow slowly) or it may converge because of cancellation among the terms. The tests which measure the first type of convergence are the most obvious and these are the "elementary" ones that we discuss in the present section.

Proposition 3.15 (The Comparison Test) *Suppose that $\sum_{j=1}^{\infty} a_j$ is a convergent series of nonnegative terms. If $\{b_j\}$ are real numbers and if $|b_j| \leq a_j$ for every j then the series $\sum_{j=1}^{\infty} b_j$ converges.*

Proof: Because the first series converges, it satisfies the Cauchy criterion for series. Hence, given $\epsilon > 0$, there is an N so large that if $n \geq m > N$ then

$$\left| \sum_{j=m}^{n} a_j \right| < \epsilon.$$

But then

$$\left| \sum_{j=m}^{n} b_j \right| \leq \sum_{j=m}^{n} |b_j| \leq \sum_{j=m}^{n} a_j < \epsilon.$$

It follows that the series $\sum b_j$ satisfies the Cauchy criterion for series. Therefore it converges. $\qquad\square$

Corollary 3.16 If $\sum_{j=1}^{\infty} a_j$ is as in the proposition and if $0 \leq b_j \leq a_j$ for every j then the series $\sum_{j=1}^{\infty} b_j$ converges.

Proof: Obvious. $\qquad\square$

EXAMPLE 3.17 The series $\sum_{j=1}^{\infty} 2^{-j} \sin j$ is seen to converge by comparing it with the series $\sum_{j=1}^{\infty} 2^{-j}$. $\qquad\blacksquare$

Theorem 3.18 (The Cauchy Condensation Test) Assume that $a_1 \geq a_2 \geq \cdots \geq a_j \geq \ldots 0$. The series

$$\sum_{j=1}^{\infty} a_j$$

converges if and only if the series

$$\sum_{k=1}^{\infty} 2^k \cdot a_{2^k}$$

converges.

Proof: First assume that the series $\sum_{j=1}^{\infty} a_j$ converges. Notice that, for each $k \geq 1$,

$$2^{k-1} \cdot a_{2^k} = \underbrace{a_{2^k} + a_{2^k} + \cdots + a_{2^k}}_{2^{k-1} \text{ times}}$$

$$\leq a_{2^{k-1}+1} + a_{2^{k-1}+2} + \cdots + a_{2^k}.$$

$$= \sum_{m=2^{k-1}+1}^{2^k} a_m$$

Therefore

$$\sum_{k=1}^{N} 2^{k-1} \cdot a_{2^k} \leq \sum_{k=1}^{N} \sum_{m=2^{k-1}+1}^{2^k} a_m = \sum_{m=2}^{2^N} a_m .$$

Since the partial sums on the right are bounded (because the series $\sum_j a_j$ converges), so are the partial sums on the left. It follows that the series

$$\sum_{k=1}^{\infty} 2^k \cdot a_{2^k} = 2 \sum_{k=1}^{\infty} 2^{k-1} \cdot a_{2^k}$$

converges.

For the converse, assume that the series

$$\sum_{k=1}^{\infty} 2^k \cdot a_{2^k} \tag{3.18.1}$$

converges. Observe that, for $k \geq 1$,

$$
\begin{aligned}
\sum_{m=2^{k-1}+1}^{2^k} a_j \;&=\; a_{2^{k-1}+1} + a_{2^{k-1}+2} + \cdots + a_{2^k} \\[2mm]
&\leq\; \underbrace{a_{2^{k-1}} + a_{2^{k-1}} + \cdots + a_{2^{k-1}}}_{2^{k-1}\ \text{times}} \\[2mm]
&=\; 2^{k-1} \cdot a_{2^{k-1}} .
\end{aligned}
$$

It follows that

$$
\begin{aligned}
\sum_{m=2}^{2^N} a_j \;&=\; \sum_{k=1}^{N} \sum_{m=2^{k-1}+1}^{2^k} a_m \\[2mm]
&\leq\; \sum_{k=1}^{N} 2^{k-1} \cdot a_{2^{k-1}} .
\end{aligned}
$$

By the hypothesis that the series (3.18.1) converges, the partial sums on the right must be bounded. But then the partial sums on the left are bounded as well. Since the summands a_j are nonnegative, the series on the left converges. $\square$

POINT OF CONFUSION 3.19 In order to apply the Cauchy condensation test correctly, and effectively, it must be that the terms of the series decrease to zero. The proof shows why this needs to be true.

EXAMPLE 3.20 We apply the Cauchy condensation test to the harmonic series

$$\sum_{j=1}^{\infty} \frac{1}{j}.$$

It leads us to examine the series

$$\sum_{k=1}^{\infty} 2^k \cdot \frac{1}{2^k} = \sum_{k=1}^{\infty} 1.$$

Since the latter series diverges, the harmonic series diverges as well. ∎

Proposition 3.21 (Geometric Series) *Let α be a real number. The series*

$$\sum_{j=0}^{\infty} \alpha^j$$

is called a geometric series. It converges if and only if $|\alpha| < 1$. In this circumstance, the sum of the series (that is, the limit of the partial sums) is $1/(1 - \alpha)$.

Proof: Let S_N denote the Nth partial sum of the geometric series. Then

$$\begin{aligned} \alpha \cdot S_N &= \alpha(1 + \alpha + \alpha^2 + \dots \alpha^N) \\ &= \alpha + \alpha^2 + \dots \alpha^{N+1}. \end{aligned}$$

It follows that $\alpha \cdot S_N$ and S_N are nearly the same: in fact

$$\alpha \cdot S_N + 1 - \alpha^{N+1} = S_N.$$

Solving this equation for the quantity S_N yields

$$S_N = \frac{1 - \alpha^{N+1}}{1 - \alpha}$$

when $\alpha \neq 1$.

If $|\alpha| < 1$ then $\alpha^{N+1} \to 0$, hence the sequence of partial sums tends to the limit $1/(1 - \alpha)$. If $|\alpha| > 1$ then α^{N+1} diverges, hence the sequence of partial sums diverges. This completes the proof for $|\alpha| \neq 1$. But the divergence in case $|\alpha| = 1$ follows because the summands will not tend to zero. □

EXAMPLE 3.22 Consider the series

$$\sum_{j=0}^{\infty} \frac{2^j}{3^j}.$$

Writing the series as

$$\sum_{j=0}^{\infty} \left(\frac{2}{3}\right)^j \, ,$$

we see that it is a geometric series. Since $|2/3| < 1$, the series converges. Its sum is 3. ∎

Corollary 3.23 *The series*

$$\sum_{j=1}^{\infty} \frac{1}{j^r}$$

converges if r is a real number that exceeds 1 and diverges otherwise.

Proof: When $r > 0$ we can apply the Cauchy Condensation Test. This leads us to examine the series

$$\sum_{k=1}^{\infty} 2^k \cdot 2^{-kr} = \sum_{k=1}^{\infty} \left(2^{1-r}\right)^k \, .$$

This last is a geometric series, with the role of α played by the quantity $\alpha = 2^{1-r}$. When $r > 1$ then $|\alpha| < 1$ so the series converges. Otherwise, by the Cauchy test, it diverges.

Later on, in Proposition 3.33, we learn the Integral Test. This gives another nice way to think about this example. □

EXAMPLE 3.24 Let us apply the Cauchy Condensation Test to the series

$$\sum_{j=1}^{\infty} \frac{1}{j(\log_2 j)^2} \, .$$

This leads us to examine the series

$$\sum_{k=1}^{\infty} 2^k \cdot \frac{1}{2^k \cdot k^2} = \sum_{k=1}^{\infty} \frac{1}{k^2} \, .$$

By the preceding corollary, this is a convergent series. So the original series converges.

Theorem 3.25 (The Root Test for Convergence) *Consider the series*

$$\sum_{j=1}^{\infty} a_j \, .$$

If

$$\limsup_{j \to \infty} |a_j|^{1/j} < 1$$

then the series converges.

Proof: Refer again to the discussion of the concept of limit superior in Chapter 2. By our hypothesis, there is a number $0 < \beta < 1$ and an integer $N > 0$ such that, for all $j > N$, it holds that

$$|a_j|^{1/j} < \beta \,.$$

In other words,

$$|a_j| < \beta^j \,.$$

Since $0 < \beta < 1$ the sum of the terms on the right constitutes a convergent geometric series. By the Comparison Test, the sum of the terms on the left converges. □

Theorem 3.26 (The Ratio Test for Convergence) *Consider a series*

$$\sum_{j=1}^{\infty} a_j \,.$$

If

$$\limsup_{j \to \infty} \left| \frac{a_{j+1}}{a_j} \right| < 1$$

then the series converges.

Proof: It is possible to supply a proof similar to that of the Root Test. We leave such a proof for the exercises, and instead supply an argument which relates the two tests in an interesting fashion.

Let

$$\lambda = \limsup_{j \to \infty} \left| \frac{a_{j+1}}{a_j} \right| < 1 \,.$$

Select a real number μ such that $\lambda < \mu < 1$. By the definition of $\limsup$, there is an N so large that, if $j > N$, then

$$\left| \frac{a_{j+1}}{a_j} \right| < \mu \,.$$

This may be rewritten as

$$|a_{j+1}| < \mu \cdot |a_j| \quad , \quad j \geq N \,.$$

Thus (much as in the proof of the Root Test) we have for $k \geq 0$ that

$$|a_{N+k}| \leq \mu \cdot |a_{N+k-1}| \leq \mu \cdot \mu \cdot |a_{N+k-2}| \leq \cdots \leq \mu^k \cdot |a_N| \,.$$

It is convenient to denote $N + k$ by $n, n \geq N$. Thus the last inequality reads

$$|a_n| < \mu^{n-N} \cdot |a_N|$$

or

$$|a_n|^{1/n} < \mu^{(n-N)/n} \cdot |a_N|^{1/n} .$$

Remembering that N has been fixed once and for all, we pass to the lim sup as $n \to \infty$. The result is

$$\limsup_{n \to \infty} |a_n|^{1/n} \leq \mu .$$

Since $\mu < 1$, we find that our series satisfies the hypotheses of the Root Test. Hence it converges. $\square$

POINT OF CONFUSION 3.27 The proof of the Ratio Test shows that *if* a series passes the Ratio Test then it passes the Root Test (the converse is not true, as you will learn in Exercise **2**). Put another way, the Root Test is a better test than the Ratio Test because it will give information whenever the Ratio Test does and also in some circumstances when the Ratio Test does not.

Why do we therefore learn the Ratio Test? The answer is that there are circumstances when the Ratio Test is easier to apply than the Root Test.

EXAMPLE 3.28 The series

$$\sum_{j=1}^{\infty} \frac{2^j}{j!}$$

is easily studied using the Ratio Test (recall that $j! \equiv j \cdot (j-1) \cdot \ldots 2 \cdot 1$). Indeed $a_j = 2^j/j!$ and

$$\left| \frac{a_{j+1}}{a_j} \right| = \frac{2^{j+1}/(j+1)!}{2^j/j!} .$$

We can perform the division to see that

$$\left| \frac{a_{j+1}}{a_j} \right| = \frac{2}{j+1} .$$

The lim sup of the last expression is 0. By the Ratio Test, the series converges.

Notice that in this example, while the Root Test applies in principle, it would be difficult to use in practice.   ∎

POINT OF CONFUSION 3.29 How do we know when to apply the Root Test and when to apply the Ratio Test? The most definitive answer to this question is that you learn from experience.

But we can give these guidelines. If the summands involve products, such as factorials, then the Ratio Test is probably most appropriate. If instead the summands involve powers, then the Root Test is probably most appropriate.

EXAMPLE 3.30 We apply the Root Test to the series

$$\sum_{j=1}^{\infty} \frac{j^2}{2^j} .$$

Observe that

$$a_j = \frac{j^2}{2^j}$$

hence that

$$|a_j|^{1/j} = \frac{\left(j^{1/j}\right)^2}{2} .$$

As $j \to \infty$, we see that

$$\limsup_{j \to \infty} |a_j|^{1/j} = \frac{1}{2} .$$

By the Root Test (refer to Example 2.40), the series converges. ∎

It is natural to ask whether the Ratio and Root Tests can detect divergence. Neither test is necessary and sufficient: there are series which elude the analysis of both tests. However, the arguments that we used to establish Theorems 3.25 and 3.26 can also be used to establish the following (the proofs are left as exercises):

Theorem 3.31 (The Root Test for Divergence) *Consider the series*

$$\sum_{j=1}^{\infty} a_j$$

of nonzero terms. If

$$\limsup_{j \to \infty} |a_j|^{1/j} > 1$$

then the series diverges.

Theorem 3.32 (The Ratio Test for Divergence) *Consider the series*

$$\sum_{j=1}^{\infty} a_j$$

of nonzero terms. If there is an $N > 0$ such that

$$\left| \frac{a_{j+1}}{a_j} \right| \geq 1 , \quad \forall j \geq N$$

then the series diverges.

In both the Root Test and the Ratio Test, if the lim sup is equal to 1, then no single conclusion is possible. The exercises give examples of series, some of which converge and some of which do not, in which these tests give lim sup equal to 1.

We conclude this section by saying a word about the integral test.

Proposition 3.33 (The Integral Test) *Let f be a continuous function on $[0, \infty)$ that is monotonically decreasing. The series*

$$\sum_{j=1}^{\infty} f(j)$$

converges if and only if the integral

$$\int_{1}^{\infty} f(x)\, dx$$

converges.

We have not treated the integral yet in this book, so we shall not prove the result here. We note that it is easy to apply the integral test to the function $f(x) = 1/x$ to see that the harmonic series diverges.

A Look Back

1. Explain the Cauchy Condensation Test in the language of comparison of series.
2. Why would the Comparison Test not be relevant to a series with both positive and negative terms?
3. What is the slowest converging series that you know?
4. What is the fastest converging series that you know?

Exercises

1. Let p be a polynomial with no constant term. If $b_j > 0$ for every j and if $\sum_{j=1}^{\infty} b_j$ converges then prove that the series $\sum_{j=1}^{\infty} p(b_j)$ converges.
2. Examine the series
   $$\frac{1}{3} + \frac{1}{5} + \frac{1}{3^2} + \frac{1}{5^2} + \frac{1}{3^3} + \frac{1}{5^3} + \frac{1}{3^4} + \frac{1}{5^4} + \cdots$$
   Prove that the Root Test shows that the series converges while the Ratio Test gives no information.
3. Check that both the Root Test and the Ratio Test give no information for the series $\sum_{j=1}^{\infty} \frac{1}{j}$, $\sum_{j=1}^{\infty} \frac{1}{j^2}$. However, one of these series is divergent and the other is convergent.

4. Prove Theorem 3.31.

5. Prove Theorem 3.32.

6. Let a_j be a sequence of real numbers. Define
$$m_j = \frac{a_1 + a_2 + \ldots a_j}{j} .$$
Prove that, if $\lim_{j \to \infty} a_j = \ell$, then $\lim_{j \to \infty} m_j = \ell$. Give an example to show that the converse is not true.

7. Imitate the proof of the Root Test to give a direct proof of the Ratio Test.

8. Let $\sum_j a_j$ and $\sum_j b_j$ be series of positive terms. Prove that if there is a constant $C > 0$ such that
$$\frac{1}{C} \leq \frac{a_j}{b_j} \leq C$$
for all j large, then either both series diverge or both series converge.

9. TRUE or FALSE: If the a_j are positive and $\sum a_j$ converges then $\sum a_j/j$ converges.

10. Use the integral test to determine whether each of these series converges:
    (a) $\sum 1/(j \log j)$
    (b) $\sum 1/j^2$
    (c) $\sum e^{-j}$
    (d) $\sum j/2^j$

11. The Ratio Test tells us nothing about the series
$$\sum_j \frac{1}{j^\alpha |\log j|^\beta}$$
for $\alpha > 0$, $\beta > 0$. Can you use some other reasoning to comment on the convergence of this series?

12. Is there any value of $\alpha > 0$ for which
$$\sum_j \frac{1}{j |\log j|^\alpha}$$
converges?

3.3 Advanced Convergence Tests

Preliminary Remarks

Whereas the previous section treated convergence tests for positive series, we now study convergence tests for series that contain both positive and negative terms. These are both more mysterious and more interesting.

The types of series studied here arise in the theory of Fourier series and in the study of solutions of partial differential equations.

In this section we consider convergence tests for series which depend on cancellation among the terms of the series. One of the most profound of these depends on a technique called *Summation by Parts*. You may wonder whether this process is at all related to the "integration by parts" procedure that you learned in calculus—it certainly has a similar form. Indeed it will turn out (and we shall see the details of this assertion as the book develops) that summing a series and performing an integration are two aspects of the same limiting process. The Summation by Parts method is merely our first glimpse of this relationship.

Proposition 3.34 (Summation by Parts) *Let $\{a_j\}_{j=0}^\infty$ and $\{b_j\}_{j=0}^\infty$ be two sequences of real numbers. For $N = 0, 1, 2, \ldots$ set*

$$A_N = \sum_{j=0}^{N} a_j$$

(we adopt the convention that $A_{-1} = 0$). Then, for any $0 \leq m \leq n < \infty$, it holds that

$$\sum_{j=m}^{n} a_j \cdot b_j = [A_n \cdot b_n - A_{m-1} \cdot b_m]$$

$$+ \sum_{j=m}^{n-1} A_j \cdot (b_j - b_{j+1}).$$

Proof: We write

$$\sum_{j=m}^{n} a_j \cdot b_j = \sum_{j=m}^{n} (A_j - A_{j-1}) \cdot b_j$$

$$= \sum_{j=m}^{n} A_j \cdot b_j - \sum_{j=m}^{n} A_{j-1} \cdot b_j$$

$$= \sum_{j=m}^{n} A_j \cdot b_j - \sum_{j=m-1}^{n-1} A_j \cdot b_{j+1}$$

$$= \sum_{j=m}^{n-1} A_j \cdot (b_j - b_{j+1}) + A_n \cdot b_n - A_{m-1} \cdot b_m.$$

This is what we wished to prove. $\square$

Now we apply Summation by Parts to prove a convergence test due to Niels Henrik Abel (1802-1829).

Theorem 3.35 (Abel's Convergence Test) *Consider the series*

$$\sum_{j=0}^{\infty} a_j \cdot b_j .$$

Suppose that

1. *The partial sums $A_N = \sum_{j=0}^{N} a_j$ form a bounded sequence;*

2. $b_0 \geq b_1 \geq b_2 \geq \ldots;$

3. $\lim_{j \to \infty} b_j = 0.$

Then the original series

$$\sum_{j=0}^{\infty} a_j \cdot b_j$$

converges.

Proof: Suppose that the partial sums A_N are bounded in absolute value by a number K. Pick $\epsilon > 0$ and choose an integer N so large that $b_N < \epsilon/(2K)$. For $N \leq m \leq n < \infty$ we use the partial summation formula to write

$$\left| \sum_{j=m}^{n} a_j \cdot b_j \right| = \left| A_n \cdot b_n - A_{m-1} \cdot b_m + \sum_{j=m}^{n-1} A_j \cdot (b_j - b_{j+1}) \right|$$

$$\leq K \cdot |b_n| + K \cdot |b_m| + K \cdot \sum_{j=m}^{n-1} |b_j - b_{j+1}| .$$

Now we take advantage of the facts that $b_j \geq 0$ for all j and that $b_j \geq b_{j+1}$ for all j to estimate the last expression by

$$K \cdot \left[b_n + b_m + \sum_{j=m}^{n-1} (b_j - b_{j+1}) \right] .$$

[Notice that the expressions $b_j - b_{j+1}, b_m$, and b_n are all nonnegative.] Now the sum collapses and the last line is estimated by

$$K \cdot [b_n + b_m - b_n + b_m] = 2 \cdot K \cdot b_m .$$

By our choice of N the right side is smaller than ϵ. Thus our series satisfies the Cauchy criterion and therefore converges. $\qquad\square$

EXAMPLE 3.36 (THE ALTERNATING SERIES TEST) As a first application of Abel's convergence test, we examine alternating series. Consider a series of the form

$$\sum_{j=1}^{\infty} (-1)^j \cdot b_j, \qquad (3.36.1)$$

with $b_1 \geq b_2 \geq b_3 \geq \cdots \geq 0$ and $b_j \to 0$ as $j \to \infty$. We set $a_j = (-1)^j$ and apply Abel's test. We see immediately that all partial sums A_N are either -1 or 0. In particular, this sequence of partial sums is bounded. And the b_j terms are decreasing and tending to zero. By Abel's convergence test, the alternating series (3.36.1) converges. ∎

Proposition 3.37 Let $b_1 \geq b_2 \geq \ldots$ and assume that $b_j \to 0$. Consider the alternating series $\sum_{j=1}^{\infty} (-1)^j b_j$ as in the last example. It is convergent: let S be its sum. Then the partial sums S_N satisfy $|S - S_N| \leq b_{N+1}$.

Proof: Observe that

$$|S - S_N| = |b_{N+1} - b_{N+2} + b_{N+3} - + \ldots|.$$

But

$$
\begin{aligned}
b_{N+2} - b_{N+3} + - \ldots &\leq b_{N+2} + (-b_{N+3} + b_{N+3}) \\
&\qquad + (-b_{N+5} + b_{N+5}) + \ldots \\
&= b_{N+2}
\end{aligned}
$$

and

$$
\begin{aligned}
b_{N+2} - b_{N+3} + - \ldots &\geq (b_{N+2} - b_{N+2}) + (b_{N+4} - b_{N+4}) + \ldots \\
&= 0.
\end{aligned}
$$

It follows that

$$|S - S_N| \leq |b_{N+1}|$$

as claimed. □

EXAMPLE 3.38 Consider the series

$$\sum_{j=1}^{\infty} (-1)^j \frac{1}{j}.$$

Then the partial sum $S_{100} = -.688172$ is within 0.01 (in fact within $1/101$) of the full sum S and the partial sum $S_{10000} = -.6930501$ is within 0.0001 (in fact within $1/10001$) of the sum S. ∎

EXAMPLE 3.39 Next we examine a series which is important in the study of Fourier analysis. Consider the series

$$\sum_{j=1}^{\infty} \frac{\sin j}{j} \, . \tag{3.39.1}$$

We already know that the series $\sum \frac{1}{j}$ diverges. However, the expression $\sin j$ changes sign in a rather sporadic fashion. We might hope that the series (3.39.1) converges because of cancellation of the summands. We take $a_j = \sin j$ and $b_j = 1/j$. Abel's test will apply if we can verify that the partial sums A_N of the a_j terms are bounded. To see this we use a trick:

Observe that

$$\cos(j + 1/2) = \cos j \cdot \cos 1/2 - \sin j \cdot \sin 1/2$$

and

$$\cos(j - 1/2) = \cos j \cdot \cos 1/2 + \sin j \cdot \sin 1/2 \, .$$

Subtracting these equations and solving for $\sin j$ yields that

$$\sin j = \frac{\cos(j - 1/2) - \cos(j + 1/2)}{2 \cdot \sin 1/2} \, .$$

We conclude that

$$A_N = \sum_{j=1}^{N} a_j = \sum_{j=1}^{N} \frac{\cos(j - 1/2) - \cos(j + 1/2)}{2 \cdot \sin 1/2} \, .$$

Of course this sum collapses and we see that

$$A_N = \frac{-\cos(N + 1/2) + \cos 1/2}{2 \cdot \sin 1/2} \, .$$

Thus

$$|A_N| \le \frac{2}{2 \cdot \sin 1/2} = \frac{1}{\sin 1/2} \, ,$$

independent of N.

Thus the hypotheses of Abel's test are verified and the series

$$\sum_{j=1}^{\infty} \frac{\sin j}{j}$$

converges.   ∎

Remark 3.40 It is interesting to notice that both the series

$$\sum_{j=1}^{\infty} \frac{|\sin j|}{j} \quad \text{and} \quad \sum_{j=1}^{\infty} \frac{\sin^2 j}{j}$$

diverge. The proofs of these assertions are left as exercises for you.

We turn next to the topic of absolute and conditional convergence. A series of real constants

$$\sum_{j=1}^{\infty} a_j$$

is said to be *absolutely convergent* if

$$\sum_{j=1}^{\infty} |a_j|$$

converges. We have:

Proposition 3.41 *If the series $\sum_{j=1}^{\infty} a_j$ is absolutely convergent, then it is convergent.*

Proof: This is an immediate corollary of the Comparison Test.   □

Definition 3.42 A series $\sum_{j=1}^{\infty} a_j$ is said to be *conditionally convergent* if $\sum_{j=1}^{\infty} a_j$ converges, but it does not converge absolutely.

We see that absolutely convergent series are convergent but the next example shows that the converse is not true.

EXAMPLE 3.43 The series

$$\sum_{j=1}^{\infty} \frac{(-1)^j}{j}$$

converges by the Alternating Series Test. However, it is not absolutely convergent because the harmonic series

$$\sum_{j=1}^{\infty} \frac{1}{j}$$

diverges.   ∎

There is a remarkable robustness result for absolutely convergent series that fails dramatically for conditionally convergent series. This result is enunciated in the next theorem. We first need a definition.

Definition 3.44 Let $\sum_{j=1}^{\infty} a_j$ be a given series. Let $\{p_j\}_{j=1}^{\infty}$ be a sequence in which every positive integer occurs once and only once (but not necessarily in the usual order). We call $\{p_j\}$ a *permutation* of the natural numbers.

Then the series

$$\sum_{j=1}^{\infty} a_{p_j}$$

is said to be a *rearrangement* of the given series.

Theorem 3.45 (Riemann, Weierstrass) *If the series $\sum_{j=1}^{\infty} a_j$ of real numbers is absolutely convergent to a (limiting) sum ℓ, then every rearrangement of the series converges also to ℓ.*

If the real series $\sum_{j=1}^{\infty} b_j$ is conditionally convergent and if β is any real number or $\pm\infty$ then there is a rearrangement of the series that converges to β.

Proof: We prove the first assertion here and explore the second in the exercises.

Let us choose a rearrangement of the given series and denote it by $\sum_{j=1}^{\infty} a_{p_j}$, where p_j is a permutation of the positive integers. Pick $\epsilon > 0$. By the hypothesis that the original series converges absolutely we may choose an integer $N > 0$ such that $N < m \leq n < \infty$ implies that

$$\sum_{j=m}^{n} |a_j| < \epsilon. \tag{3.45.1}$$

[The presence of the absolute values in the left side of this inequality will prove crucial in a moment.] Choose a positive integer M such that $M \geq N$ and the integers $1, \ldots, N$ are all contained in the list $p_1, p_2, \ldots, p_M$. If $K > M$ then the partial sum $\sum_{j=1}^{K} a_j$ will trivially contain the summands $a_1, a_2, \ldots a_N$. Also the partial sum $\sum_{j=1}^{K} a_{p_j}$ will contain the summands $a_1, a_2, \ldots a_N$. It follows that

$$\sum_{j=1}^{K} a_j - \sum_{j=1}^{K} a_{p_j}$$

will contain only summands *after* the Nth one in the original series. By inequality (3.45.1) we may conclude that

$$\left| \sum_{j=1}^{K} a_j - \sum_{j=1}^{K} a_{p_j} \right| \leq \sum_{j=N+1}^{\infty} |a_j| \leq \epsilon.$$

We conclude that the rearranged series converges; and it converges to the same sum as the original series. $\square$

A Look Back

1. What does Summation by Parts say? How does it work?

2. How is Summation by Parts similar to integration by parts?

3. What is Abel's convergence test?

4. Why does an alternating series converge?

Exercises

1. If $1/2 > b_j > 0$ for every j and if $\sum_{j=1}^{\infty} b_j$ converges then prove that $\sum_{j=1}^{\infty} \frac{b_j}{1-b_j}$ converges.

2. Follow these steps to give another proof of the Alternating Series Test: **a)** Prove that the odd partial sums form an increasing sequence; **b)** Prove that the even partial sums form a decreasing sequence; **c)** Prove that every even partial sum majorizes all subsequent odd partial sums; **d)** Use a pinching principle.

3. If $b_j > 0$ and $\sum_{j=1}^{\infty} b_j$ converges then prove that

$$\sum_{j=1}^{\infty} (b_j)^{1/2} \cdot \frac{1}{j^{\alpha}}$$

converges for any $\alpha > 1/2$. Give an example to show that the assertion is false if $\alpha = 1/2$.

4. Let p be a polynomial with integer coefficients and degree at least 1. Let $b_1 \geq b_2 \geq \cdots \geq 0$ and assume that $b_j \rightarrow 0$. Prove that if $(-1)^{p(j)}$ is not always positive and not always negative then in fact it will alternate in sign so that $\sum_{j=1}^{\infty} (-1)^{p(j)} \cdot b_j$ will converge.

5. Use Abel's test to see that a series of the form

$$\sum_{j=1}^{\infty} (-1)^{3j} a_j ,$$

with the a_j positive numbers tending monotonically to zero, converges.

6. Apply summation by parts to the series

$$\sum_{j=1}^{\infty} j 2^{-j} .$$

What can you say about the sum of this series?

* 7. Assume that $\sum_{j=1}^{\infty} b_j$ is a convergent series of positive real numbers. Let $s_j = \sum_{\ell=1}^{j} b_\ell$. Discuss convergence or divergence for the series $\sum_{j=1}^{\infty} s_j \cdot b_j$. Discuss convergence or divergence for the series $\sum_{j=1}^{\infty} \frac{b_j}{1+s_j}$.

* **8.** If $b_j > 0$ for every j and if $\sum_{j=1}^{\infty} b_j$ diverges then define $s_j = \sum_{\ell=1}^{j} b_\ell$. Discuss convergence or divergence for the series $\sum_{j=1}^{\infty} \frac{b_j}{s_j}$.

* **9.** Let $\sum_{j=1}^{\infty} b_j$ be a conditionally convergent series of real numbers. Let β be a real number. Prove that there is a rearrangement of the series that converges to β. (**Hint:** First observe that the positive terms of the given series must form a divergent series. Also, the negative terms form a divergent series. Now build the rearrangement by choosing finitely many positive terms whose sum "just exceeds" β. Then add on enough negative terms so that the sum is "just less than" β. Repeat this oscillatory procedure.)

* **10.** Prove that

$$\sum_{j=1}^{\infty} \frac{|\sin j|}{j} \quad \text{and} \quad \sum_{j=1}^{\infty} \frac{\sin^2 j}{j}$$

are both divergent series.

* **11.** What can you say about the convergence or divergence of

$$\sum_{j=1}^{\infty} \frac{(2j+3)^{1/2} - (2j)^{1/2}}{j^{3/4}} \, ?$$

* **12.** Find a rearrangement of the series $\sum_j (-1)^j / j$ that converges to 10. Find a rearrangement that converges to π.

3.4 Some Special Series

Preliminary Remarks

When we studied sequences we collected a library of basic sequences to which we could compare other sequences. Just so, in the study of series we want some basic series to which we can compare new series. That is what we shall address in the present section.

We begin with a series that defines a special constant of mathematical analysis.

Definition 3.46 The series

$$\sum_{j=0}^{\infty} \frac{1}{j!},$$

where $j! \equiv j \cdot (j-1) \cdot (j-2) \cdots 1$ for $j \geq 1$ and $0! \equiv 1$, is convergent (by the Ratio Test, for instance). Its sum is denoted by the symbol e in honor of the Swiss mathematician Léonard Euler, who first studied it (see also Example

2.44, where the number e is studied by way of a sequence). We shall see in Proposition 3.47 that these two approaches to the number e are equivalent.

Like the number π, to be considered later in this book, the number e is one which arises repeatedly in a number of contexts in mathematics. It has many special properties. We first relate the series definition of e to the sequence definition:

Proposition 3.47 *The limit*

$$\lim_{n \to \infty} \left(1 + \frac{1}{n}\right)^n$$

exists and equals e.

Proof: We need to compare the quantities

$$A_N \equiv \sum_{j=0}^{N} \frac{1}{j!} \quad \text{and} \quad B_N \equiv \left(1 + \frac{1}{N}\right)^N .$$

We use the binomial theorem to expand B_N :

$$
\begin{aligned}
B_N &= 1 + \frac{N}{1} \cdot \frac{1}{N} + \frac{N \cdot (N-1)}{2 \cdot 1} \cdot \frac{1}{N^2} + \frac{N \cdot (N-1) \cdot (N-2)}{3 \cdot 2 \cdot 1} \cdot \frac{1}{N^3} \\
&\quad + \cdots \frac{N}{1} \cdot \frac{1}{N^{N-1}} + 1 \cdot \frac{1}{N^N} \\
&= 1 + 1 + \frac{1}{2!} \cdot \frac{N-1}{N} + \frac{1}{3!} \cdot \frac{N-1}{N} \cdot \frac{N-2}{N} + \cdots \\
&\quad + \frac{1}{(N-1)!} \cdot \frac{N-1}{N} \cdot \frac{N-2}{N} \cdots \frac{2}{N} \\
&\quad + \frac{1}{N!} \cdot \frac{N-1}{N} \cdot \frac{N-2}{N} \cdots \frac{1}{N} \\
&= 1 + 1 + \frac{1}{2!} \cdot \left(1 - \frac{1}{N}\right) + \frac{1}{3!} \cdot \left(1 - \frac{1}{N}\right) \cdot \left(1 - \frac{2}{N}\right) + \cdots \\
&\quad + \frac{1}{(N-1)!} \cdot \left(1 - \frac{1}{N}\right) \cdot \left(1 - \frac{2}{N}\right) \cdots \left(1 - \frac{N-2}{N}\right) \\
&\quad + \frac{1}{N!} \cdot \left(1 - \frac{1}{N}\right) \cdot \left(1 - \frac{2}{N}\right) \cdots \left(1 - \frac{N-1}{N}\right) .
\end{aligned}
$$

Notice that every summand that appears in this last equation is positive. Thus, for $0 \le M \le N$,

$$B_N \geq 1 + 1 + \frac{1}{2!} \cdot \left(1 - \frac{1}{N}\right) + \frac{1}{3!} \cdot \left(1 - \frac{1}{N}\right) \cdot \left(1 - \frac{2}{N}\right)$$
$$+ \cdots + \frac{1}{M!} \cdot \left(1 - \frac{1}{N}\right) \left(1 - \frac{2}{N}\right) \cdots \left(1 - \frac{M-1}{N}\right) .$$

In this last inequality we hold M fixed and let N tend to infinity. The result is that

$$\liminf_{N \to \infty} B_N > 1 + 1 + \frac{1}{2!} + \frac{1}{3!} + \cdots + \frac{1}{M!} = A_M .$$

Now, as $M \to \infty$, the quantity A_M converges to e (by the *definition* of e). So we obtain

$$\liminf_{N \to \infty} B_N \geq e . \tag{3.47.1}$$

On the other hand, our expansion for B_N allows us to observe that $B_N \leq A_N$. Thus

$$\limsup_{N \to \infty} B_N \leq e . \tag{3.47.2}$$

Combining (3.47.1) and (3.47.2), we find that

$$e \leq \liminf_{N \to \infty} B_N \leq \limsup_{N \to \infty} B_N \leq e$$

hence that $\lim_{N \to \infty} B_N$ exists and equals e. This is the desired result. □

Remark 3.48 The last proof illustrates the value of the concepts of lim inf and lim sup. For we do not know in advance that the limit of the expressions B_N exists, much less that the limit equals e. However, the lim inf and the lim sup always exist. So we estimate those instead, and find that they are equal and that they equal e.

A Look Back

1. What is the series representation for e^x?

2. What is the series representation for $\sin x$?

3. Why is the power series representation for e^x unique?

4. Give another series representation for e^x besides the power series representation.

Exercises

1. Prove a formula for the sum of the first N perfect squares.

2. A real number s is called *algebraic* if it satisfies a polynomial equation of the form

$$a_0 + a_1 x + a_2 x^2 + \cdots + a_m x^m = 0$$

   with the coefficients a_j being integers and $a_m \neq 0$. Prove that, if we replace the word "integers" in this definition with "rational numbers," then the set of algebraic numbers remains the same. Prove that $n^{p/q}$ is algebraic for any positive integers p, q, n. A number which is not algebraic is called *transcendental*.

3. Discuss convergence of $\sum_j 1/[\ln(j+1)]^k$ for k a positive integer.

4. Refer to Example 3.39 and Remark 3.40. What can you say about the convergence of $\sum_j [\sin j]^k / j$ for k a positive integer?

5. Discuss convergence of $\sum_j 1/p(j)$ for p a polynomial.

6. Discuss convergence of $\sum_j \exp(p(j))$ for p a polynomial.

7. Give a series expansion for $\ln n$.

* 8. At least one of the numbers $e + \pi$ and $e - \pi$ is transcendental. Explain why.

* 9. Refer to Exercise **2** for terminology. A number is said to be *transcendental* if it is not algebraic. It is quite difficult to give an explicit example of a transcendental number. However, use a counting argument to show that transcendental numbers exist.

10. Discuss convergence of the series

$$\sum_j \frac{2^j}{j!} \, .$$

11. Give sufficient conditions on $b_j > 0$ to guarantee that

$$\sum_j \frac{b_j}{j}$$

   converges.

* 12. Refer to Exercise **2** for terminology. A real number is called *transcendental* if it is not algebraic. Prove that the number of algebraic numbers is countable. Explain why this implies that the number of transcendental numbers is uncountable. Thus most real numbers are transcendental; however, it is extremely difficult to verify that any particular real number is transcendental. It is known that π and e are transcendental, but the proofs are extremely difficult.

3.5   Operations on Series

Preliminary Remarks

Since series are algebraic objects, it is natural that we would want to perform arithmetic operations on them. In the present section we examine these operations, and see how series behave in this context.

Some operations on series, such as addition, subtraction, and scalar multiplication, are straightforward. Others, such as multiplication, entail subtleties. This section treats all these matters.

Proposition 3.49 *Let*

$$\sum_{j=1}^{\infty} a_j \quad and \quad \sum_{j=1}^{\infty} b_j$$

be convergent series of real numbers; assume that the series sum to limits α and β respectively. Then

(a) *The series $\sum_{j=1}^{\infty}(a_j + b_j)$ converges to the limit $\alpha + \beta$.*

(b) *If c is a constant then the series $\sum_{j=1}^{\infty} c \cdot a_j$ converges to $c \cdot \alpha$.*

Proof: We shall prove assertion **(a)** and leave the easier assertion **(b)** as an exercise.

Pick $\epsilon > 0$. Choose an integer N_1 so large that $n > N_1$ implies that the partial sum $S_n \equiv \sum_{j=1}^{n} a_j$ satisfies $|S_n - \alpha| < \epsilon/2$. Choose N_2 so large that $n > N_2$ implies that the partial sum $T_n \equiv \sum_{j=1}^{n} b_j$ satisfies $|T_n - \beta| < \epsilon/2$. If U_n is the nth partial sum of the series $\sum_{j=1}^{\infty}(a_j + b_j)$ and if $n > N_0 \equiv \max(N_1, N_2)$ then

$$|U_n - (\alpha + \beta)| \leq |S_n - \alpha| + |T_n - \beta| < \frac{\epsilon}{2} + \frac{\epsilon}{2} = \epsilon.$$

Thus the sequence $\{U_n\}$ converges to $\alpha + \beta$. This proves part **(a)**. The proof of **(b)** is similar.   □

Of course subtraction of series is covered by part **(a)** of the last proposition.

In order to keep our discussion of multiplication of series as straightforward as possible, we deal at first with absolutely convergent series. It is convenient in this discussion to begin our sums at $j = 0$ instead of $j = 1$. If we wish to multiply

$$\sum_{j=0}^{\infty} a_j \quad \text{and} \quad \sum_{j=0}^{\infty} b_j \, ,$$

then we need to specify what the partial sums of the product series should be. An obvious necessary condition that we wish to impose is that, if the first series converges to α and the second converges to β, then the product series, whatever we define it to be, should converge to $\alpha \cdot \beta$.

The naive method for defining the summands of the product series $\sum_j c_j$ is to let $c_j = a_j \cdot b_j$. However, a glance at the product of two partial sums of the given series shows that such a definition would be ignoring the distributivity of multiplication over addition.

Cauchy's idea was that the summands for the product series should be

$$c_n \equiv \sum_{j=0}^{n} a_j \cdot b_{n-j} \, .$$

This particular form for the summands can be easily motivated using power series considerations. For now we concentrate on verifying that this "Cauchy product" of two series really works.

Theorem 3.50 *Let $\sum_{j=0}^{\infty} a_j$ and $\sum_{j=0}^{\infty} b_j$ be two absolutely convergent series which converge to limits α and β respectively. Define the series $\sum_{m=0}^{\infty} c_m$ with summands $c_m = \sum_{j=0}^{m} a_j \cdot b_{m-j}$. Then the series $\sum_{m=0}^{\infty} c_m$ converges to $\alpha \cdot \beta$.*

Proof: Let A_n, B_n, and C_n be the partial sums of the three series in question. We calculate that

$$\begin{aligned} C_n &= (a_0 b_0) + (a_0 b_1 + a_1 b_0) + (a_0 b_2 + a_1 b_1 + a_2 b_0) \\ &\quad + \cdots + (a_0 b_n + a_1 b_{n-1} + \cdots + a_n b_0) \\ &= a_0 \cdot B_n + a_1 \cdot B_{n-1} + a_2 \cdot B_{n-2} + \cdots + a_n \cdot B_0 \, . \end{aligned}$$

We set $\lambda_n = B_n - \beta$, each n, and rewrite the last line as

$$\begin{aligned} C_n &= a_0(\beta + \lambda_n) + a_1(\beta + \lambda_{n-1}) + \cdots a_n(\beta + \lambda_0) \\ &= A_n \cdot \beta + [a_0 \lambda_n + a_1 \cdot \lambda_{n-1} + \cdots + a_n \cdot \lambda_0] \, . \end{aligned}$$

Denote the expression in square brackets by the symbol ρ_n. Suppose that we could show that $\lim_{n \to \infty} \rho_n = 0$. Then we would have

$$\begin{aligned} \lim_{n \to \infty} C_n &= \lim_{n \to \infty} (A_n \cdot \beta + \rho_n) \\ &= \left(\lim_{n \to \infty} A_n \right) \cdot \beta + \left(\lim_{n \to \infty} \rho_n \right) \\ &= \alpha \cdot \beta + 0 \\ &= \alpha \cdot \beta \, . \end{aligned}$$

Thus it is enough to examine the limit of the expressions ρ_n.

Since $\sum_{j=1}^{\infty} a_j$ is absolutely convergent, we know that $A = \sum_{j=1}^{\infty} |a_j|$ is a finite number. Choose $\epsilon > 0$. Since $\sum_{j=1}^{\infty} b_j$ converges to β it follows that $\lambda_n \to 0$. Thus we may choose an integer $N > 0$ such that $n > N$ implies that $|\lambda_n| < \epsilon$. Thus, for $n = N + k, k > 0$, we may estimate

$$
\begin{aligned}
|\rho_{N+k}| \;\leq\; & |\lambda_0 a_{N+k} + \lambda_1 a_{N+k-1} + + \cdots + \lambda_N a_k| \\
& + |\lambda_{N+1} a_{k-1} + \lambda_{N+2} a_{k-2} + \cdots + \lambda_{N+k} a_0| \\
\leq\; & |\lambda_0 a_{N+k} + \lambda_1 a_{N+k-1} + + \cdots + \lambda_N a_k| \\
& + \max_{p\geq 1}\{|\lambda_{N+p}|\} \cdot (|a_{k-1}| + |a_{k-2}| + \cdots + |a_0|) \\
\leq\; & (N+1) \cdot \max_{\ell \geq k} |a_\ell| \cdot \max_{0 \leq j \leq N} |\lambda_j| + \epsilon \cdot A .
\end{aligned}
$$

With N fixed, we let $k \to \infty$ in the last inequality. Since $\max_{\ell \geq k} |a_\ell| \to 0$, we find that

$$\limsup_{n \to \infty} |\rho_n| \leq \epsilon \cdot A .$$

Since $\epsilon > 0$ was arbitrary, we conclude that

$$\lim_{n \to \infty} |\rho_n| \to 0 .$$

This completes the proof. □

POINT OF CONFUSION 3.51 The idea of the product of series is sophisticated. But, if we keep in mind how we multiply polynomials, then the ideas will fall into place. Namely,

$$
\begin{aligned}
& (a_0 + a_1 x + a_2 x^2 + a_3 x^3 + \cdots + a_k x^k) \cdot (b_0 + b_1 x + b_2 x^2 + b_3 x^3 + \cdots + a_k x_k) \\
= \; & a_0 b_0 + (a_0 b_1 + a_1 b_0)x + (a_0 b_2 + a_1 b_1 + a_2 b_0)x^2 + (a_0 b_3 + a_1 b_2 + a_2 b_1 + a_3 b_0)x^3 \\
& + \cdots + (a_0 b_k + a_1 b_{k-1} + a_2 b_{k-2} + \cdots + a_k b_0)x^k + \cdots .
\end{aligned}
$$

And we recognize the coefficients c_n as the coefficients of this product.

The main point to remember is that we do *not* multiply series termwise.

Notice that, in the proof of the theorem, we really only used the fact that one of the given series was absolutely convergent, not that both were absolutely convergent. Some hypothesis of this nature is necessary, as the following example shows.

EXAMPLE 3.52 Consider the Cauchy product of the two conditionally convergent series

$$\sum_{j=0}^{\infty} \frac{(-1)^j}{\sqrt{j+1}} \quad \text{and} \quad \sum_{j=0}^{\infty} \frac{(-1)^j}{\sqrt{j+1}} .$$

Observe that

$$
\begin{aligned}
c_m &= \frac{(-1)^0(-1)^m}{\sqrt{1}\sqrt{m+1}} + \frac{(-1)^1(-1)^{m-1}}{\sqrt{2}\sqrt{m}} + \cdots \\
&\quad + \frac{(-1)^m(-1)^0}{\sqrt{m+1}\sqrt{1}} \\
&= \sum_{j=0}^{m} (-1)^m \frac{1}{\sqrt{(j+1) \cdot (m+1-j)}} .
\end{aligned}
$$

However, for $0 \le j \le m$,

$$
(j+1) \cdot (m+1-j) \le (m+1) \cdot (m+1) = (m+1)^2 .
$$

Thus

$$
|c_m| \ge \sum_{j=0}^{m} \frac{1}{m+1} = 1 .
$$

We thus see that the terms of the series $\sum_{m=0}^{\infty} c_m$ do not tend to zero, so the series cannot converge. ∎

A Look Back

1. Suppose that $\sum_j a_j$ and $\sum_j b_j$ are absolutely convergent series. Can you make sense of the Cauchy product of $\sum_j a_j^2$ and $\sum_j b_j^2$?

2. Where does the hypothesis of absolute convergence come into play in our argument for the validity of the Cauchy product?

3. Does the Cauchy product distribute over addition?

4. Is the Cauchy product commutative?

Exercises

1. Let $\sum_{j=1}^{\infty} a_j$ and $\sum_{j=1}^{\infty} b_j$ be convergent series of positive real numbers. Discuss convergence of $\sum_{j=1}^{\infty} a_j b_j$.

2. Prove Proposition 3.49(b).

3. Calculate the Cauchy product of the series $\sum_j 1/j^2$ and the series $\sum_j 1/j^4$.

4. Look up the definition of "ring" on Google. Prove that the set of all absolutely convergent series forms a ring.

5. Motivate the form of the Cauchy product using multiplication of polynomials.

6. Calculate the Cauchy product of $\sum_j x^j$ and $\sum_j x^{2j}$.

7. How many different divergent series are there? How many different absolutely convergent series are there?

8. Do there exist terms $a_j > 0$ so that $\sum_j a_j$ converges and also $\sum_j \log a_j$ converges?

9. Show that it is never the case that if $a_j > 0$ and $\sum a_j$ converges then $\sum_j e^{a_j}$ converges.

10. TRUE or FALSE: If $a_j > 0$ and $\sum a_j$ converges then $\sum \sin a_j$ converges.

* 11. Discuss the concept of the exponential of a power series.

* 12. Let $\sum_{j=1}^{\infty} a_j$ and $\sum_{j=1}^{\infty} b_j$ be convergent series of positive real numbers. Discuss division of these two series. Use the idea of the Cauchy product.

Chapter 4

Basic Topology

4.1 Open and Closed Sets

Preliminary Remarks

Topology is, in a sense, a generalization of classical Euclidean geometry. But, where classical geometry studies rigid equivalences of triangles and rectangles, topology studies more flexible equivalences of a great variety of shapes. The subject of topology as we know it today was founded largely by Henri Poincaré. Over the course of the twentieth century it has developed into a lively and intensely studied discipline. This section is your introduction to the subject.

To specify a topology on a set is to describe certain subsets that will play the role of neighborhoods. These are called *open sets*.

In what follows, we will use "interval notation": If $a \leq b$ are real numbers then we define

$$
\begin{aligned}
(a,b) &= \{x \in \mathbb{R} : a < x < b\}, \\
[a,b] &= \{x \in \mathbb{R} : a \leq x \leq b\}, \\
[a,b) &= \{x \in \mathbb{R} : a \leq x < b\}, \\
(a,b] &= \{x \in \mathbb{R} : a < x \leq b\}.
\end{aligned}
$$

Intervals of the form (a,b) are called *open*. Those of the form $[a,b]$ are called *closed*. The other two are termed *half-open* or *half-closed*. See Figure 4.1.

Now we extend the terms "open" and "closed" to more general sets.

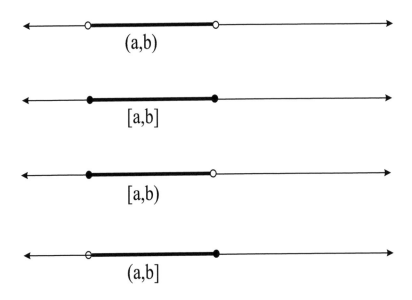

Figure 4.1: Intervals.

Definition 4.1 A set $U \subseteq \mathbb{R}$ is called *open* if, for each $x \in U$, there is an $\epsilon > 0$ such that the interval $(x - \epsilon, x + \epsilon)$ is contained in U. See Figure 4.2.

Remark 4.2 The interval $(x - \epsilon, x + \epsilon)$ is frequently termed a *neighborhood* of x, and is commonly denoted $N_\epsilon(x)$.

EXAMPLE 4.3 The set $U = \{x \in \mathbb{R} : |x - 3| < 2\}$ is open. To see this, choose a point $x \in U$. Let $\epsilon = 2 - |x - 3| > 0$. Then we claim that the interval $I = (x - \epsilon, x + \epsilon) \subseteq U$.
    For, if $t \in I$, then

$$
\begin{aligned}
|t - 3| &= |(t - x) + (x - 3)| \\
&\leq |t - x| + |x - 3| \\
&< \epsilon + |x - 3| \\
&= (2 - |x - 3|) + |x - 3| \\
&= 2 .
\end{aligned}
$$

But this means that $t \in U$.
    We have shown that $t \in I$ implies $t \in U$. Therefore $I \subseteq U$. It follows from the definition that U is open. ∎

POINT OF CONFUSION 4.4 The way to think about the definition of open set is that a set is open when none of its elements is at the "edge" of the set—each

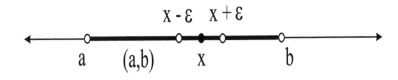

Figure 4.2: An open set.

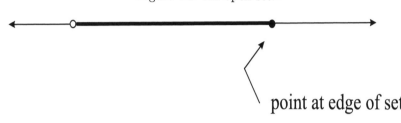

point at edge of set

Figure 4.3: A set that is not open.

element is surrounded by other elements of the set, indeed a whole interval of them. See Figure 4.3. The remainder of this section will make these comments precise.

Proposition 4.5 *If U_α are open sets, for α in some (possibly uncountable) index set A, then*

$$U = \bigcup_{\alpha \in A} U_\alpha$$

is open.

Proof: Let $x \in U$. By definition of union, the point x must lie in some U_α. But U_α is open. Therefore there is an interval $I = (x - \epsilon, x + \epsilon)$ such that $I \subseteq U_\alpha$. Therefore certainly $I \subseteq U$. This proves that U is open. □

Proposition 4.6 *If $U_1, U_2, \ldots, U_k$ are open sets then the set*

$$V = \bigcap_{j=1}^{k} U_j$$

is also open.

Proof: Let $x \in V$. Then $x \in U_j$ for each j. Since each U_j is open there is for each j a positive number ϵ_j such that $I_j = (x - \epsilon_j, x + \epsilon_j)$ lies in U_j. Set $\epsilon = \min\{\epsilon_1, \ldots, \epsilon_k\}$. Then $\epsilon > 0$ and $(x - \epsilon, x + \epsilon) \subseteq I_j \subseteq U_j$ for every j. But

Figure 4.4: Structure of an open set.

that just means that $(x - \epsilon, x + \epsilon) \subseteq V$. Therefore V is open.   □

Notice the difference between these two propositions: arbitrary unions of open sets are open. But, in order to guarantee that an intersection of open sets is still open, we had to assume that we were only intersecting finitely many such sets. To understand this matter, bear in mind the example of the open sets

$$U_j = \left(-\frac{1}{j}, \frac{1}{j}\right) \;, \quad j = 1, 2, \dots .$$

The intersection of the open sets U_j is the singleton $\{0\}$, which is not open.

POINT OF CONFUSION 4.7 It is natural to think of an open set as an interval without endpoints—what we usually call an open interval. And that is not far from the whole truth. We shall prove below that absolutely any open set is the disjoint union of at most countably many open intervals.

The same analysis as in the first example shows that, if $a < b$, then the interval (a, b) is an open set. On the other hand, intervals of the form $(a, b]$ or $[a, b)$ or $[a, b]$ are *not* open. In the first instance, the point b is the center of no interval $(b - \epsilon, b + \epsilon)$ contained in $(a, b]$. Think about the other two intervals to understand why they are not open. We call intervals of the form (a, b) *open intervals*.

We are now in a position to give a complete description of all open sets.

Proposition 4.8 *Let $U \subseteq \mathbb{R}$ be a nonempty open set. Then there are either finitely many or countably many pairwise disjoint open intervals I_j such that*

$$U = \bigcup_{j=1}^{\infty} I_j \, .$$

See Figure 4.4.

Proof: Assume that U is an open subset of the real line. We define an equivalence relation on the set U. The resulting equivalence classes (see Appendix II) will be the open intervals I_j.

Let a and b be elements of U. We say that a is related to b if all real numbers between a and b are also elements of U. It is obvious that this relation is both reflexive and symmetric. For transitivity notice that if a is

Figure 4.5: A closed set.

related to b and b is related to c then (assuming that a, b, c are distinct) one of the numbers a, b, c must lie between the other two. Assume for simplicity that $a < b < c$. Then all numbers between a and c lie in U, for all such numbers are either between a and b or between b and c or are b itself. Thus a is related to c. (The other possible orderings of a, b, c are left for you to consider.)

Thus we have an equivalence relation on the set U. Call the equivalence classes $\{U_\alpha\}_{\alpha \in A}$. We claim that each U_α is an open interval. In fact if a, b are elements of some U_α then all points between a and b are in U. But then a moment's thought shows that each of those "in between" points is related to both a and b. Therefore all points between a and b are elements of U_α. We conclude that U_α is an interval. Is it an *open* interval?

Let $x \in U_\alpha$. Then $x \in U$ so that there is an open interval $I = (x - \epsilon, x + \epsilon)$ contained in U. But x is related to all the elements of I; it follows that $I \subseteq U_\alpha$. Therefore U_α is open.

We have exhibited the set U as a union of open intervals. These intervals are pairwise disjoint because they arise as the equivalence classes of an equivalence relation. Finally, each of these open intervals contains a (different) rational number (why?). Therefore there can be at most countably many of the intervals U_α.   □

Definition 4.9 A subset $F \subseteq \mathbb{R}$ is called *closed* if the complement $\mathbb{R} \setminus F$ is open. See Figure 4.5.

EXAMPLE 4.10 The set $[0, 1]$ is closed. For its complement is

$$(-\infty, 0) \cup (1, \infty),$$

which is certainly open.   ∎

EXAMPLE 4.11 An interval of the form $[a, b] = \{x : a \le x \le b\}$ is closed. For its complement is $(-\infty, a) \cup (b, \infty)$, which is the union of two open intervals.

The finite set $A = \{-4, -2, 5, 13\}$ is closed because its complement is

$$(-\infty, -4) \cup (-4, -2) \cup (-2, 5) \cup (5, 13) \cup (13, \infty),$$

which is open.

The set $B = \{1, 1/2, 1/3, 1/4, \ldots\} \cup \{0\}$ is closed, for its complement is the set

$$(-\infty, 0) \cup \left\{ \bigcup_{j=1}^{\infty} (1/(j+1), 1/j) \right\} \cup (1, \infty),$$

which is open.

Verify for yourself that if the point 0 is omitted from the set B, then the set is no longer closed. ∎

POINT OF CONFUSION 4.12 A very common point of confusion among beginners is to think that *any* set is either open or closed. This is simply not true. For example, the set $S = [0, 1)$ is not open (because it contains its left endpoint) and not closed (because it does not contain its right endpoint). The fact is that most sets of reals are *neither* open nor closed.

Proposition 4.13 *If E_α are closed sets, for α in some (possibly uncountable) index set A, then*

$$E = \bigcap_{\alpha \in A} E_\alpha$$

is closed.

Proof: This is just the contrapositive of Proposition 4.5 above: if U_α is the complement of E_α, each α, then U_α is open. Then $U = \cup U_\alpha$ is also open. But then

$$E = \bigcap E_\alpha = \bigcap {}^c (U_\alpha) = {}^c \left(\bigcup U_\alpha \right) = {}^c U$$

is closed. Here ${}^c S$ denotes the complement of a set S. □

The fact that the set B in the last example is closed, but that $B \setminus \{0\}$ is not, is placed in perspective by the next proposition.

Proposition 4.14 *Let S be a set of real numbers. Then S is closed if and only if every Cauchy sequence $\{s_j\}$ of elements of S has a limit which is also an element of S.*

Proof: First suppose that S is closed and let $\{s_j\}$ be a Cauchy sequence in S. We know, since the reals are complete, that there is an element $s \in \mathbb{R}$ such that $s_j \to s$. The point of this half of the proof is to see that $s \in S$. If this statement were false then $s \in U = \mathbb{R} \setminus S$. But U must be open since it is the complement of a closed set. Thus there is an $\epsilon > 0$ such that the interval $I = (s - \epsilon, s + \epsilon) \subseteq U$. This means that no element of S lies in I. In particular, $|s - s_j| \geq \epsilon$ for every j. This contradicts the statement that $s_j \to s$. We conclude that $s \in S$.

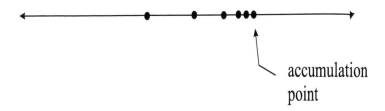

Figure 4.6: The idea of an accumulation point.

Conversely, assume that every Cauchy sequence in S has its limit in S. If S were not closed then its complement would not be open. Hence there would be a point $t \in \mathbb{R} \setminus S$ with the property that no interval $(t - \epsilon, t + \epsilon)$ lies in $\mathbb{R} \setminus S$. In other words, $(t - \epsilon, t + \epsilon) \cap S \neq \emptyset$ for every $\epsilon > 0$. Thus for $j = 1, 2, 3, \ldots$ we may choose a point $s_j \in (t - 1/j, t + 1/j) \cap S$. It follows that $\{s_j\}$ is a sequence of elements of S that converge to $t \in \mathbb{R} \setminus S$. That contradicts our hypothesis. We conclude that S must be closed.   □

Let S be a subset of $\mathbb{R}$. A point x is called an *accumulation point* of S if every neighborhood of x contains infinitely many distinct elements of S. See Figure 4.6. In particular, x is an accumulation point of S if it is the limit of a sequence of distinct elements in S. The last proposition tells us that closed sets are characterized by the property that they contain all of their accumulation points.

A Look Back

1. Give a verbal description of an open set.

2. Give a verbal description of a closed set.

3. Give an example of a set that is neither open nor closed.

4. What is an accumulation point?

Exercises

1. Let S be any set and $\epsilon > 0$. Define $T = \{t \in \mathbb{R} : |t - s| < \epsilon \text{ for some } s \in S\}$. Prove that T is open.

2. Give an example of nonempty *closed* sets $X_1 \supseteq X_2 \supseteq \ldots$ such that $\cap_j X_j = \emptyset$.

3. Give an example of nonempty *closed* sets $X_1 \subseteq X_2 \ldots$ such that $\cup_j X_j$ is open.

4. Give an example of open sets $U_1 \supseteq U_2 \ldots$ such that $\cap_j U_j$ is closed and nonempty.

5. Exhibit a countable collection of open sets U_j such that each open set $\mathcal{O} \subseteq \mathbb{R}$ can be written as a union of some of the sets U_j.

6. Let S be any closed set and define, for $x \in \mathbb{R}$,

$$\mathrm{dist}(x, S) = \inf\{|x - s| : s \in S\}\,.$$

Prove that, if $x \notin S$, then $\mathrm{dist}(x, S) > 0$. If $x, y \in \mathbb{R}$ then prove that

$$|\mathrm{dis}(x, S) - \mathrm{dis}(y, S)| \le |x - y|\,.$$

7. Let S be a set of real numbers. If S is not open then must it be closed? If S is not closed then must it be open?

8. Let E be a closed set and F a closed and bounded set. Assume that E and F are disjoint. Show that there is an $\epsilon > 0$ so that

$$|e - f| > \epsilon$$

for all $e \in E$ and $f \in F$.

9. Show that the conclusion of Exercise **8** is false if E and F are both closed but not bounded.

* 10. The *closure* of a set S is the intersection of all closed sets that contain S. We denote the closure of S by $\overline{S}$. Call a set S *robust* if it is the closure of its interior (where the *interior* of S is the set of all $x \in S$ so that there is an $\epsilon > 0$ with $(x - \epsilon, x + \epsilon) \subseteq S$). Which sets of reals are robust?

* 11. Let S be an uncountable subset of $\mathbb{R}$. Prove that S must have infinitely many accumulation points. Must it have uncountably many?

* 12. Let S be any set and define $V = \{t \in \mathbb{R} : |t - s| \le 1 \text{ for some } s \in S\}$. Is V necessarily closed?

4.2   Further Properties of Open and Closed Sets

Preliminary Remarks

Open and closed sets are the basic elements of topology. Understanding the topology of a space consists in understanding its open and closed sets. In this section we move beyond the basics and explore some of the deeper properties of open and closed sets.

Let $S \subseteq \mathbb{R}$ be a set. We call $b \in \mathbb{R}$ a *boundary point* of S if every nonempty neighborhood $(b - \epsilon, b + \epsilon)$ contains both points of S and points of $\mathbb{R} \setminus S$. See Figure 4.7. We denote the set of boundary points of S by ∂S.

A boundary point b might lie in S and might lie in the complement of S. The next example serves to illustrate the concept:

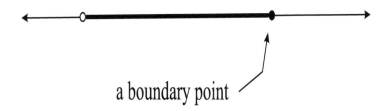

a boundary point

Figure 4.7: The idea of a boundary point.

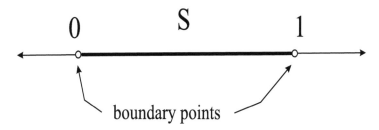

0   S   1

boundary points

Figure 4.8: Boundary of the open unit interval.

EXAMPLE 4.15 Let S be the interval $(0,1)$. Then no point of $(0,1)$ is in the boundary of S since every point of $(0,1)$ has a neighborhood that lies entirely inside $(0,1)$ (in other words, every point is an interior point—see Exercise **10** of the last section). Also, no point of the complement of $T = [0,1]$ lies in the boundary of T for a similar reason.

Indeed, the only candidates for elements of the boundary of S are 0 and 1. See Figure 4.8. The point 0 *is* an element of the boundary since every neighborhood $(0 - \epsilon, 0 + \epsilon)$ contains the point $\epsilon/2 \in S$ and the point $-\epsilon/2 \in \mathbb{R} \setminus S$. A similar calculation shows that 1 lies in the boundary of S.

Now consider the set $T = [0,1]$. Certainly there are no boundary points in $(0,1)$, for the same reason as in the first paragraph. And there are no boundary points in $\mathbb{R} \setminus [0,1]$, since that set is open. Thus the only candidates for elements of the boundary are 0 and 1. As in the first paragraph, these are both indeed boundary points for T. See Figure 4.9.

Notice that neither of the boundary points of S lie in S while both of the boundary points of T lie in T. ∎

EXAMPLE 4.16 The boundary of the set $\mathbb{Q} \subseteq \mathbb{R}$ is the entire real line. For if x is any element of $\mathbb{R}$ then every interval $(x - \epsilon, x + \epsilon)$ contains both rational numbers and irrational numbers. ∎

The union of a set S with its boundary is the *closure* of S, denoted $\overline{S}$ (Exercise **10** at the end of the last section discusses this idea from a different point of view). The next example illustrates the concept.

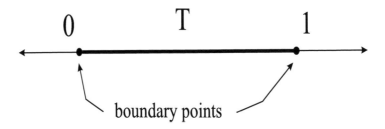

Figure 4.9: Boundary of the closed unit interval.

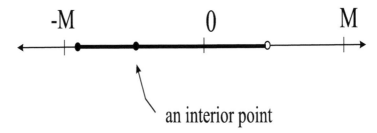

Figure 4.10: The idea of an interior point.

EXAMPLE 4.17 Let S be the set of rational numbers in the interval $[0, 1]$.
Then the closure $\overline{S}$ of S is the entire interval $[0, 1]$.

Let T be the open interval $(0, 1)$. Then the closure $\overline{T}$ of T is the closed interval $[0, 1]$.   ∎

Definition 4.18 Let $S \subseteq \mathbb{R}$. A point $s \in S$ is called an *interior point* of S if there is an $\epsilon > 0$ such that the interval $(s - \epsilon, s + \epsilon)$ lies in S. See Figure 4.10. We call the set of all interior points the *interior* of S, and we denote this set by $\overset{\circ}{S}$.

A point $t \in S$ is called an *isolated point* of S if there is an $\epsilon > 0$ such that the intersection of the interval $(t - \epsilon, t + \epsilon)$ with S is just the singleton $\{t\}$. See Figure 4.11.

By the definitions given here, an isolated point t of a set $S \subseteq \mathbb{R}$ is a boundary point. For any interval $(t - \epsilon, t + \epsilon)$ contains a point of S (namely,

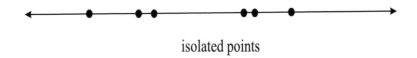

isolated points

Figure 4.11: The idea of an isolated point.

t itself) and points of $\mathbb{R} \setminus S$ (since t is isolated).

Proposition 4.19 *Let $S \subseteq \mathbb{R}$. Then each point of S is either an interior point or a boundary point of S.*

Proof: Fix $s \in S$. If s is not an interior point then no open interval centered at s contains only elements of s. Thus any interval centered at s contains an element of S (namely, s itself) and also contains points of $\mathbb{R} \setminus S$. Thus s is a boundary point of S. □

POINT OF CONFUSION 4.20 Let $S = \mathbb{Q}$. Then the interior of S is empty and the boundary of S is all of $\mathbb{R}$.

By contrast, let $S = \mathbb{R}$. Then the interior of S is all of $\mathbb{R}$ and the boundary of S is empty.

EXAMPLE 4.21 Let $S = [0, 1]$. Then the interior points of S are the elements of $(0, 1)$. The boundary points of S are the points 0 and 1. The set S has no isolated points.

Let $T = \{1, 1/2, 1/3, \ldots\} \cup \{0\}$. Then the points 1, 1/2, 1/3, ... are isolated points of T. The point 0 is an accumulation point of T. Every element of T is a boundary point, and there are no others. ∎

Remark 4.22 Observe that the interior points of a set S are *elements* of S— by their very definition. Also isolated points of S are elements of S. However, a boundary point of S may or may not be an element of S.

If x is an accumulation point of S then every open neighborhood of x contains infinitely many elements of S. Hence x is either a boundary point of S or an interior point of S; it *cannot* be an isolated point of S.

Proposition 4.23 *Let S be a subset of the real numbers. Then the boundary of S equals the boundary of $\mathbb{R} \setminus S$.*

Proof: Exercise. □

The next theorem allows us to use the concept of boundary to distinguish open sets from closed sets.

Theorem 4.24 *A closed set contains all of its boundary points. An open set contains none of its boundary points.*

Proof: Let S be closed and let x be an element of its boundary. If every neighborhood of x contains points of S *other than x itself* then x is an accumulation point of S hence $x \in S$. If not every neighborhood of x

Figure 4.12: A bounded set.

contains points of S other than x itself, then there is an $\epsilon > 0$ such that $\{(x - \epsilon, x) \cup (x, x + \epsilon)\} \cap S = \emptyset$. The only way that x can be an element of ∂S in this circumstance is if $x \in S$. That is what we wished to prove.

For the other half of the theorem notice that if T is open then ^cT is closed. But then ^cT will contain all its boundary points, which are the same as the boundary points of T itself (why is this true?). Thus T can contain none of its boundary points.   □

Proposition 4.25 *Every nonisolated boundary point of a set S is an accumulation point of the set S.*

Proof: This proof is treated in the exercises.   □

Definition 4.26 A subset S of the real numbers is called *bounded* if there is a positive number M such that $|s| \le M$ for every element s of S. See Figure 4.12.

The next result is one of the great theorems of nineteenth century analysis. It is essentially a restatement of the Bolzano-Weierstrass theorem of Section 2.2.

Theorem 4.27 (Bolzano-Weierstrass) *Every bounded, infinite subset of $\mathbb{R}$ has an accumulation point.*

Proof: Let S be a bounded, infinite set of real numbers. Let $\{a_j\}$ be a sequence of distinct elements of S. By Theorem 2.24, there is a subsequence $\{a_{j_k}\}$ that converges to a limit α. Then α is an accumulation point of S. □

Corollary 4.28 *Let $S \subseteq \mathbb{R}$ be a nonempty, closed, and bounded set. If $\{a_j\}$ is any sequence in S, then there is a Cauchy subsequence $\{a_{j_k}\}$ that converges to an element of S.*

Proof: Merely combine the Bolzano-Weierstrass theorem with Proposition 4.14 of the last section.   □

A Look Back

1. Say in words what an interior point is.
2. Say in words what a boundary point is.
3. Say in words what an isolated point is.
4. Explain why an isolated point is always a boundary point.

Exercises

1. Let S be any set of real numbers. Prove that $S \subseteq \overline{S}$. Prove that $\overline{S}$ is a closed set. Prove that $\overline{S} \setminus \overset{\circ}{S}$ is the boundary of S.

2. Determine the interior and the boundary of the Cantor set.

3. Prove Proposition 4.23.

4. The union of infinitely many closed sets need not be closed. It need not be open either. Give examples to illustrate the possibilities.

5. The intersection of infinitely many open sets need not be open. It need not be closed either. Give examples to illustrate the possibilities.

6. Give an example of a one-to-one, onto, continuous function f with a continuous inverse from the halfline $(0, \infty)$ to the full line $(-\infty, \infty)$.

7. Prove Proposition 4.25.

8. Let S be any set of real numbers. Prove that $\overset{\circ}{S}$ is open. Prove that S is open if and only if S equals its interior.

9. Let U be a closed set in the plane. Show that its projection on the x-axis is not necessarily closed.

10. Let $E \subseteq \mathbb{R}$ be closed. Let U be the complement of E. Prove that U is the countable union of open intervals.

11. What is the distance of the Cantor set to the point $1/2$.

4.3 Compact Sets

Preliminary Remarks
Compact sets are a relatively recent development in mathematics. A compact set is an infinite set that, in certain key ways, behaves like a finite set. While this may sound like a conundrum, the proof is in the pudding. This section will introduce you to a remarkable and fascinating collection of sets of real numbers.

Compact sets are sets (usually infinite) which share many of the most important properties of finite sets. They play an important role in real analysis.

Definition 4.29 A set $S \subseteq \mathbb{R}$ is called *compact* if every sequence in S has a subsequence that converges *to an element of S.*

Theorem 4.30 (Heine-Borel) *A set $S \subseteq \mathbb{R}$ is compact if and only if it is closed and bounded.*

Proof: That a closed, bounded set has the property of compactness is the content of Corollary 4.28.

Now let S be a set that is compact. If S is not bounded, then there is an element s_1 of S that has absolute value larger than 1. Also there must be an element s_2 of S that has absolute value larger than $2 + |s_1|$. Continuing, we find elements $s_j \in S$ satisfying

$$|s_j| > j + |s_{j-1}|$$

for each j. But then no subsequence of the sequence $\{s_j\}$ can be Cauchy. This contradiction shows that S must be bounded.

If S is compact but S is not closed, then there is a point x which is the limit of a sequence $\{s_j\} \subseteq S$ but which is not itself in S. But every sequence in S is, by definition of "compact," supposed to have a subsequence converging *to an element of S.* For the sequence $\{s_j\}$ that we are considering, x is the only possibility for the limit of a subsequence. Thus it must be that $x \in S$. That contradiction establishes that S is closed. $\square$

EXAMPLE 4.31 The last theorem makes it particularly easy to identify compact sets. The set $[0, 1]$ is closed and bounded, hence compact. The set $\{0, 2, 4, 8\}$ is closed and bounded, hence compact.

By contrast, the set $[0, 1)$ is bounded but *not* closed. So it is not compact.

EXAMPLE 4.32 If $A \subseteq B$ and both sets are nonempty then $A \cap B = A \neq \emptyset$. A similar assertion holds when intersecting *finitely many* nonempty sets $A_1 \supseteq A_2 \supseteq \cdots \supseteq A_k$; it holds in this circumstance that $\cap_{j=1}^k A_j = A_k$.

However, it is possible to have infinitely many nonempty nested sets with null intersection. An example is the sets $I_j = (0, 1/j)$. Certainly $I_j \supseteq I_{j+1}$ for all j yet

$$\bigcap_{j=1}^{\infty} I_j = \emptyset .$$

By contrast, if we take $K_j = [0, 1/j]$ then

$$\bigcap_{j=1}^{\infty} K_j = \{0\}.$$

The next proposition shows that compact sets have the intuitively appealing property of the collection of sets K_j rather than the unsettling property of the collection of sets I_j. ∎

Proposition 4.33 *Let*

$$K_1 \supseteq K_2 \supseteq \cdots \supseteq K_j \supseteq \ldots$$

be nonempty compact sets of real numbers. Set

$$\mathcal{K} = \bigcap_{j=1}^{\infty} K_j.$$

Then $\mathcal{K}$ is compact and $\mathcal{K} \neq \emptyset$.

Proof: Each K_j is closed and bounded hence $\mathcal{K}$ is closed and bounded. Thus $\mathcal{K}$ is compact. Let $x_j \in K_j$, each j. Then $\{x_j\} \subseteq K_1$. By compactness, there is a convergent subsequence $\{x_{j_k}\}$ with limit $x_0 \in K_1$. However, $\{x_{j_k}\}_{k=2}^{\infty} \subseteq K_2$. Thus $x_0 \in K_2$. Similar reasoning shows that $x_0 \in K_m$ for all $m = 1, 2, \ldots$. In conclusion, $x_0 \in \cap_j K_j = \mathcal{K}$. □

A Look Back

1. Describe in words what a compact set is.

2. What does the complement of a compact set look like? Is it open? Is it bounded?

3. Is a finite set compact?

4. Is the intersection of two compact sets compact? Is the union of two compact sets compact?

Exercises

1. Let K be a compact set and let U be an open set that contains K. Prove that there is an $\epsilon > 0$ such that, if $x \in K$, then the interval $(x - \epsilon, x + \epsilon)$ is contained in U.

2. Let K be a compact set. Let $\mathcal{U} = \{U_j\}_{j=1}^k$ be a finite covering of K by open sets. Show that there is a $\delta > 0$ so that, if x is any point of K, then the disc or interval of center x and radius δ lies entirely in one of the U_j.

3. Prove that the intersection of a compact set and a closed set is compact.

4. Assume that we have intervals $[a_1, b_1] \supseteq [a_2, b_2] \supseteq \cdots$ and that $\lim_{j \to \infty} |a_j - b_j| = 0$. Prove that there is a point x such that $x \in [a_j, b_j]$ for every j.

5. If K in $\mathbb{R}$ is compact then show that ^cK is not compact.

6. Prove that the intersection of any number of compact sets is compact. The analogous statement for unions is false.

7. Let $U \subset \mathbb{R}$ be any open set. Show that there exist compact sets $K_1 \subset K_2 \subset \cdots$ so that $\cup_j K_j = U$.

8. Produce an open set U in the real line so that U may *not* be written as the decreasing intersection of compact sets.

9. Prove that the union of finitely many compact sets is compact.

10. Prove that the union of countably many compact sets is not necessarily compact.

* 11. Let K be a compact set. Let $\delta > 0$. Prove that there is a finite collection of intervals of radius δ that covers K.

4.4   The Cantor Set

Preliminary Remarks

Certainly one of the most amazing and mysterious sets ever constructed is the Cantor set. While elementary to define, the Cantor set has a fractal-like character and offers many mysteries. It is one hundred years old, but is still intensely studied today.

In this section we describe the construction of a remarkable subset of $\mathbb{R}$ with many pathological properties. It only begins to suggest the richness of the structure of the real number system.

We begin with the unit interval $S_0 = [0, 1]$. We extract from S_0 its open middle third; thus $S_1 = S_0 \setminus (1/3, 2/3)$. Observe that S_1 consists of two closed intervals of equal length $1/3$. See Figure 4.13.

Now we construct S_2 from S_1 by extracting from each of its two intervals the middle third: $S_2 = [0, 1/9] \cup [2/9, 3/9] \cup [6/9, 7/9] \cup [8/9, 1]$. Figure 4.14 shows S_2.

Continuing in this fashion, we construct S_{j+1} from S_j by extracting the middle third from each of its component subintervals. We define the Cantor

0   1

Figure 4.13: Construction of the Cantor set.

0   1

Figure 4.14: Second step in the construction of the Cantor set.

set C to be

$$C = \bigcap_{j=1}^{\infty} S_j \, .$$

Notice that each of the sets S_j is closed and bounded, hence compact. By Proposition 4.33 of the last section, C is therefore not empty. The set C is closed and bounded, hence compact.

Proposition 4.34 *The Cantor set C has zero length, in the sense that the complementary set $[0, 1] \setminus C$ has length 1.*

Proof: In the construction of S_1, we removed from the unit interval one interval of length 3^{-1}. In constructing S_2, we further removed two intervals of length 3^{-2}. In constructing S_j, we removed 2^{j-1} intervals of length 3^{-j}. Thus the total length of the intervals removed from the unit interval is

$$\sum_{j=1}^{\infty} 2^{j-1} \cdot 3^{-j} \, .$$

This last equals

$$\frac{1}{3} \sum_{j=0}^{\infty} \left(\frac{2}{3} \right)^j \, .$$

The geometric series sums easily and we find that the total length of the intervals removed is

$$\frac{1}{3} \left(\frac{1}{1 - 2/3} \right) = 1 \, .$$

Thus the Cantor set has length zero because its complement in the unit interval has length one.   □

POINT OF CONFUSION 4.35 The Cantor set is uncountable, as we shall see below. It contains no intervals. In fact it is disconnected in a very strong sense to be discussed below. Only countably many of the elements of the Cantor set are endpoints of the component intervals. Uncountably many of the points are non-endpoints.

Proposition 4.36 *The Cantor set is uncountable.*

Proof: We assign to each element of the Cantor set a "label" consisting of a sequence of 0s and 1s that identifies its location in the set.

Fix an element x in the Cantor set. Then certainly x is in S_1. If x is in the left half of S_1, then the first digit in the "label" of x is 0; otherwise it is 1. Likewise $x \in S_2$. By the first part of this argument, it is either in the left half S_{21} of S_2 (when the first digit in the label is 0) or the right half S_{22} of S_2 (when the first digit of the label is 1). Whichever of these is correct, that half will consist of two intervals of length 3^{-2}. If x is in the leftmost of these two intervals then the second digit of the "label" of x is 0. Otherwise the second digit is 1. Continuing in this fashion, we may assign to x an infinite sequence of 0s and 1s.

Conversely, if $a, b, c, \ldots$ is a sequence of 0s and 1s, then we may locate a unique corresponding element y of the Cantor set. If the first digit is a zero then y is in the left half of S_1; otherwise y is in the right half of S_1. Likewise the second digit locates y within S_2, and so forth.

Thus we have a one-to-one correspondence between the Cantor set and the collection of all infinite sequences of zeroes and ones. [Notice that we are in effect thinking of the point assigned to a sequence $c_1 c_2 c_3 \ldots$ of 0s and 1s as the limit of the points assigned to $c_1, c_1 c_2, c_1 c_2 c_3, \ldots$ Thus we are using the fact that C is closed.] However, as we can learn in Appendix II, Section A2.7, at the end of the book, the set of all infinite sequences of zeroes and ones is uncountable. Thus we see that the Cantor set is uncountable. □

Remark 4.37 A useful way to think about the Cantor set is in terms of series. Namely, C is the set of all numbers between 0 and 1 inclusive which can be written in the form

$$\sum_{j=1}^{\infty} \frac{a_j}{3^j}. \tag{4.37.1}$$

where each a_j is either 0 or 2. This representation is simply an interpretation of the labeling that we used in the last proof. We invite the reader to write out some expressions like (1) (only finite sums, of course), just to see what elements of the Cantor set arise. As you read the proof of the next theorem, you should think about it in terms of this series representation.

The Cantor set is quite thin (it has zero length) but it is large in the sense that it has uncountably many elements. Also it is compact. The next result reveals a surprising, and not generally well known, property of this "thin" set.

Theorem 4.38 *Let C be the Cantor set and define*

$$S = \{x + y : x \in C, y \in C\}.$$

Then $S = [0, 2]$.

Proof: We sketch the proof here and treat the details in the exercises.

Since $C \subseteq [0, 1]$ it is clear that $S \subseteq [0, 2]$. For the reverse inclusion, fix an element $t \in [0, 2]$. Our job is to find two elements c and d in C such that $c + d = t$.

First observe that $\{x + y : x \in S_1, y \in S_1\} = [0, 2]$. Therefore there exist $x_1 \in S_1$ and $y_1 \in S_1$ such that $x_1 + y_1 = t$.

Similarly, $\{x + y : x \in S_2, y \in S_2\} = [0, 2]$. Therefore there exist $x_2 \in S_2$ and $y_2 \in S_2$ such that $x_2 + y_2 = t$.

Continuing in this fashion we may find for each j numbers x_j and y_j such that $x_j, y_j \in S_j$ and $x_j + y_j = t$. Of course $\{x_j\} \subseteq C$ and $\{y_j\} \subseteq C$ hence there are subsequences $\{x_{j_k}\}$ and $\{y_{j_k}\}$ which converge to real numbers c and d respectively. Since C is compact, we can be sure that $c \in C$ and $d \in C$. But the operation of addition respects limits, thus we may pass to the limit as $k \to \infty$ in the equation

$$x_{j_k} + y_{j_k} = t$$

to obtain

$$c + d = t.$$

Therefore $[0, 2] \subseteq \{x + y : x \in C\}$. This completes the proof. □

In the exercises at the end of the section we shall explore constructions of other Cantor sets, some of which have zero length and some of which have positive length. The Cantor set that we have discussed in detail in the present section is sometimes distinguished with the name "the Cantor ternary set." We shall also consider in the exercises other ways to construct the Cantor ternary set.

Observe that, whereas any open set is the countable or finite disjoint union of open intervals, the existence of the Cantor set shows us that there is no such structure theorem for closed sets. That is to say, we cannot hope to write an arbitrary closed set as the disjoint union of closed intervals. [However, de Morgan's Law shows that an arbitrary closed set can be written as the countable intersection of sets, each of which is the union of two disjoint closed intervals.] In fact closed intervals are atypically simple when considered as examples of closed sets.

A Look Back

1. Why is the Cantor set uncountable?

2. Why is the Cantor set compact?

3. How many connected components does the complement of the Cantor set have? [Here a "connected component" is a maximal connected piece of the complement.]

4. What is the boundary of the Cantor set?

Exercises

1. What is the interior of the Cantor set?

2. Fix the sequence $a_j = 3^{-j}, j = 1, 2, \ldots$. Consider the set S of all sums

$$\sum_{j=1}^{\infty} \mu_j a_j \, ,$$

where each μ_j is one of the numbers 0 or 2. Show that S is the Cantor set. If s is an element of S, $s = \sum \mu_j a_j$, and if $\mu_j = 0$ for all j sufficiently large, then show that s is an endpoint of one of the intervals in one of the sets S_j that were used to construct the Cantor set in the text.

3. Construct a Cantor-like set by removing the middle *fifth* from the unit interval, removing the middle fifth of each of the remaining intervals, and so on. What is the length of the set that you construct in this fashion? Is it uncountable? Is it perfect (see Section 4.6)? Is it different from the Cantor set constructed in the text?

4. Refer to Exercise **3**. Construct a Cantor set by removing, at the jth step, a middle subinterval of length 3^{-2j+1} from each existing interval. The Cantor-like set that results should have positive length. What is that length? Does this Cantor set have the other properties of the Cantor set constructed in the text?

5. Describe how to produce a two-dimensional Cantor-like set in the plane.

6. How many endpoints of intervals are there in the Cantor set? How many non-endpoints?

7. How many points in the Cantor set have finite ternary expansions? How many have infinite ternary expansions?

* 8. Let $0 < \lambda < 1$. Imitate the construction of the Cantor set to produce a perfect subset (see Section 4.6) of the unit interval whose complement has length λ.

* 9. Discuss which sequences a_j of positive numbers could be used as in Exercise **1** to construct sets which are like the Cantor set.

* 10. Let us examine the proof that $\{x + y : x \in C, y \in C\}$ equals $[0, 2]$ more carefully:

a disconnected set

Figure 4.15: The idea of disconnected.

a) Prove for each j that $\{x + y : x \in S_j, y \in S_j\}$ equals the interval $[0, 2]$.

b) For $t \in C$, explain how the subsequences $\{x_{j_k}\}$ and $\{y_{j_k}\}$ can be chosen to satisfy $x_{j_k} + y_{j_k} = t$. Observe that it is important for the proof that the index j_k be the same for both subsequences.

c) Formulate a suitable statement concerning the assertion that the binary operation of addition "respects limits" as required in the argument in the text. Prove this statement and explain how it allows us to pass to the limit in the equation $x_{j_k} + y_{j_k} = t$.

* 11. Use the characterization of the Cantor set from Exercise 1 to give a new proof of the fact that $\{x + y : x \in C, y \in C\}$ equals the interval $[0, 2]$.

4.5 Connected and Disconnected Sets

Preliminary Remarks

We assign various attributes to sets in order to help us understand their shape and form. One of those attributes is connectedness. Connected sets are very natural objects for us to study, and we want to see how they behave under the action of continuous functions. That is the purpose of our study in this section.

Let S be a set of real numbers. We say that S is *disconnected* if it is possible to find a pair of open sets U and V such that

$$U \cap S \neq \emptyset, V \cap S \neq \emptyset,$$

$$(U \cap S) \cap (V \cap S) = \emptyset,$$

and

$$S = (U \cap S) \cup (V \cap S).$$

See Figure 4.15. If no such U and V exist then we call S *connected*.

Figure 4.16: A closed interval is connected.

EXAMPLE 4.39 The set $T = \{x \in \mathbb{R} : |x| < 1, x \neq 0\}$ is disconnected. For take $U = \{x : x < 0\}$ and $V = \{x : x > 0\}$. Then

$$U \cap T = \{x : -1 < x < 0\} \neq \emptyset$$

and

$$V \cap T = \{x : 0 < x < 1\} \neq \emptyset.$$

Also $(U \cap T) \cap (V \cap T) = \emptyset$. Clearly $T = (U \cap T) \cup (V \cap T)$, hence T is disconnected. ∎

EXAMPLE 4.40 The set $X = [-1, 1]$ is connected. To see this, suppose to the contrary that there exist open sets U and V such that $U \cap X \neq \emptyset, V \cap X \neq \emptyset, (U \cap X) \cap (V \cap X) = \emptyset$, and

$$X = (U \cap X) \cup (V \cap X).$$

Choose $a \in U \cap X$ and $b \in V \cap X$. We may assume that $a < b$. Set

$$\alpha = \sup (U \cap [a, b]).$$

Now $[a, b] \subseteq X$ hence $U \cap [a, b]$ is disjoint from V. Thus $\alpha \leq b$. But ^cV is closed hence $\alpha \notin V$. It follows that $\alpha < b$.

If $\alpha \in U$ then, because U is open, there exists an $\tilde{\alpha} \in U$ such that $\alpha < \tilde{\alpha} < b$. The existence of $\tilde{\alpha}$ contradicts the definition of α as the supremum of $U \cap [a, b]$. So $\alpha \notin U$. But $\alpha \notin U$ and $\alpha \notin V$ means $\alpha \notin X$. On the other other hand, α is the supremum of a subset of X (since $a \in X, b \in X$, and X is an interval). Since X is a closed interval, we conclude that $\alpha \in X$. This contradiction shows that X must be connected. ∎

POINT OF CONFUSION 4.41 As we shall see in detail below, the connected subsets of the real numbers are the intervals. There are no other connected subsets. The disconnected sets are much more profuse, and much more varied.

With small modifications, the discussion in the last example demonstrates that any closed interval is connected (Exercise 1). See Figure 4.16. Also (see Exercise 2), we may similarly see that any open interval or half-open interval is connected. In fact the converse is true as well:

Theorem 4.42 A subset S of $\mathbb{R}$ is connected if and only if S is an interval.

Proof: We have already noted that an interval is connected.

For the converse, note that if S is not an interval, then there exist $a \in S, b \in S$ and a point t between a and b such that $t \notin S$. Define $U = \{x \in \mathbb{R} : x < t\}$ and $V = \{x \in \mathbb{R} : t < x\}$. Then U and V are open and disjoint, $U \cap S \neq \emptyset$, $V \cap S \neq \emptyset$, and

$$S = (U \cap S) \cup (V \cap S) \ .$$

Thus S is disconnected.

We have proved the contrapositive of the statement for this direction of the theorem, hence we are finished. □

The Cantor set is not connected; indeed it is disconnected in a special sense. Call a set S *totally disconnected* if, for each distinct $x \in S$, $y \in S$, there exist disjoint open sets U and V such that $x \in U, y \in V$, and $S = (U \cap S) \cup (V \cap S)$.

Proposition 4.43 *The Cantor set is totally disconnected.*

Proof: Let $x, y \in C$ be distinct and assume that $x < y$. Set $\delta = |x - y|$. Choose j so large that $3^{-j} < \delta$. Then $x, y \in S_j$, but x and y cannot both be in the same interval of S_j (since the intervals will have length equal to 3^{-j}). It follows that there is a point t between x and y that is not an element of S_j, hence certainly not an element of C. Set $U = \{s : s < t\}$ and $V = \{s : s > t\}$. Then $x \in U \cap C$ hence $U \cap C \neq \emptyset$; likewise $V \cap C \neq \emptyset$. Also $(U \cap C) \cap (V \cap C) = \emptyset$. Finally $C = (C \cap U) \cup (C \cap V)$. Thus C is totally disconnected. □

A Look Back

1. Describe verbally what a connected set is.

2. Describe verbally what a disconnected set is.

3. What is the simplest thing you can do to a connected set to make it disconnected?

4. The entire real line is connected, but the real line with one point removed is disconnected. Explain.

Exercises

1. Imitate the example in the text to prove that any closed interval is connected.

2. Imitate the example in the text to prove that any open interval or half-open interval is connected.

3. Give an example of a totally disconnected set $S \subseteq [0, 1]$ such that $\overline{S} = [0, 1]$.

4. If A is connected and B is connected then will $A \cap B$ be connected?

5. If A is connected and B is connected then will $A \cup B$ be connected?

6. If A is connected and B is disconnected then what can you say about $A \cap B$?

7. If sets U_j form the basis of a topology on a space X (that is to say, each open set in X can be written as a union of some of the U_j) and if each U_j is connected, then what can you say about X?

8. Is the set-theoretic difference of connected sets connected?

9. Prove that the union of two connected sets is connected provided that the two sets have at least one point in common.

* **10.** Let $S \subseteq \mathbb{R}$ be a set. Let $s, t \in S$. We say that s and t are in the same *connected component* of S if the entire interval $[s, t]$ lies in S. What are the connected components of the Cantor set? Is it possible to have a set S with countably many connected components? With uncountably many connected components?

* **11.** Write the real line as the union of two totally disconnected sets.

* **12.** If A is connected and B is connected then does it follow that $A \times B$ is connected?

4.6   Perfect Sets

Preliminary Remarks

A perfect set is a very special type of set that makes it better than closed, better than compact, and remarkable in a number of ways. The Cantor set is perfect. There are many examples of perfect sets. We learn their lore in this section.

A set $S \subseteq \mathbb{R}$ is called *perfect* if it is closed and if every point of S is an accumulation point of S. The property of being perfect is a rather special one: it implies that the set has no isolated points.

Obviously a closed interval $[a, b]$ is perfect. After all, a point x in the interior of the interval is surrounded by an entire open interval $(x - \epsilon, x + \epsilon)$ of elements of the interval; moreover a is the limit of elements from the right and b is the limit of elements from the left.

Perhaps more surprising is that the Cantor set, *a totally disconnected set*, is perfect. It is certainly closed. Now fix $x \in C$. Then certainly $x \in S_1$. Thus x is in one of the two intervals composing S_1. One (or perhaps both) of the endpoints of that interval does not equal x. Call that endpoint a_1. Likewise

$x \in S_2$. Therefore x lies in one of the intervals of S_2. Choose an endpoint a_2 of that interval which does not equal x. Continuing in this fashion, we construct a sequence $\{a_j\}$. Notice that *each of the elements of this sequence lies in the Cantor set* (why?). Finally, $|x - a_j| \leq 3^{-j}$ for each j. Therefore x is the limit of the sequence. We have thus proved that the Cantor set is perfect.

The fundamental theorem about perfect sets tells us that such a set must be rather large. We have

Theorem 4.44 *A nonempty perfect set must be uncountable.*

Proof: Let S be a nonempty perfect set. Since S has accumulation points, it cannot be finite. Therefore it is either countable or uncountable.

Seeking a contradiction, we suppose that S is countable. Write $S = \{s_1, s_2, \ldots\}$. Set $U_1 = (s_1 - 1, s_1 + 1)$. Then U_1 is a neighborhood of s_1. Now s_1 is a limit point of S so there must be infinitely many elements of S lying in U_1. We select a bounded open interval U_2 such that $\overline{U}_2 \subseteq U_1$, $\overline{U}_2$ does not contain s_1, and U_2 *does* contain some element of S.

Continuing in this fashion, assume that $s_1, \ldots, s_j$ have been selected and choose a bounded interval U_{j+1} such that (i) $\overline{U}_{j+1} \subseteq U_j$, (ii) $s_j \notin \overline{U}_{j+1}$, and (iii) U_{j+1} contains some element of S.

Observe that each set $V_j = \overline{U}_j \cap S$ is closed and bounded, hence compact. Also each V_j is nonempty by construction but V_j does not contain s_{j-1}. It follows that $V \equiv \cap_j V_j$ cannot contain s_1 (since V_2 does not), cannot contain s_2 (since V_3 does not), indeed cannot contain any element of S. Hence V, being a subset of S, is empty. But V is the decreasing intersection of nonempty compact sets, hence cannot be empty!

This contradiction shows that S cannot be countable. So it must be uncountable. □

Corollary 4.45 *If $a < b$ then the closed interval $[a, b]$ is uncountable.*

Proof: The interval $[a, b]$ is perfect. □

We also have a new way of seeing that the Cantor set is uncountable, since it is perfect:

Corollary 4.46 *The Cantor set is uncountable.*

Proof: The Cantor set is nonempty and perfect. □

A Look Back

1. Describe verbally what a perfect set is.

2. What characteristics does the complement of a perfect set have (see Exercise **9** below)?

3. Explain why the rational numbers $\mathbb{Q}$ do not form a perfect set.

4. Explain why the real numbers $\mathbb{R}$ *do* form a perfect set.

Exercises

1. Let $U_1 \subseteq U_2 \ldots$ be open sets and assume that each of these sets has bounded, nonempty complement. Can it be that $\cup_j U_j = \mathbb{R}$?

2. Let $X_1, X_2, \ldots$ each be perfect sets and suppose that $X_1 \supseteq X_2 \supseteq \ldots$. Set $X = \cap_j X_j$. Is X perfect?

3. Is the product of perfect sets perfect?

4. If $A \cap B$ is perfect, then what may we conclude about A and B?

5. If $A \cup B$ is perfect, then what may we conclude about A and B?

6. Call a set imperfect if its complement is perfect. Which sets are imperfect? Can you specify a connected imperfect set?

7. A Cantor set is formed by removing not middle thirds but rather middle ninths. Which properties of the ternary Cantor set will this new set have? What will be the length of this new set?

*   8. Let $S_1, S_2, \ldots$ be closed sets and assume that $\cup_j S_j = \mathbb{R}$. Prove that at least one of the sets S_j has nonempty interior. (**Hint:** Use an idea from the proof that perfect sets are uncountable.)

*   9. Let S be a nonempty set of real numbers. A point x is called a *condensation point* of S if every neighborhood of x contains uncountably many points of S. Prove that the set of condensation points of S is closed. Is it necessarily nonempty? Is it nonempty when S is uncountable?

    If T is an uncountable set then show that the set of its condensation points is perfect.

*   10. Prove that any closed set can be written as the union of a perfect set and a countable set. (**Hint:** Refer to Exercise **9**.)

*   11. Let $0 < \alpha < 1$. Construct a Cantor-like set that has length α. Verify that this set has all the properties (except the length property) of the Cantor set that was discussed in the text.

Chapter 5

Limits and Continuity of Functions

5.1 Definition and Basic Properties of the Limit of a Function

---

Preliminary Remarks

Questions about limits go back to the ancient Greeks. The Greeks really did not understand limits (witness Zeno's paradoxes). The question of limits arose even more intensely in the development of calculus. Isaac Newton did not understand limits, and neither did Leibniz. It took the combined efforts of a number of nineteenth-century mathematical geniuses—including Cauchy, Riemann, Dirichlet, Weierstrass, and others—to finally nail down the concept of limit. Here we present the fruits of their efforts.

In this chapter we are going to treat some topics that you have seen before in your calculus class. However, we shall use the deep properties of the real numbers that we have developed in this text to obtain important new insights. Therefore you should *not* think of this chapter as review. Look at the concepts introduced here with the power of your new understanding of analysis.

Definition 5.1 Let f be a real-valued function whose domain E contains adjoining intervals (a, c) and (c, b). Let ℓ be a real number. We say that

$$\lim_{x \to c} f(x) = \ell$$

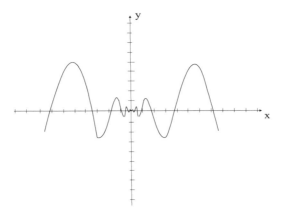

Figure 5.1: The limit of an oscillatory function.

if, for each $\epsilon > 0$, there is a $\delta > 0$ such that, when $0 < |x - c| < \delta$, then

$$|f(x) - \ell| < \epsilon.$$

POINT OF CONFUSION 5.2 The definition makes precise the notion that we can force $f(x)$ to be just as close as we please to ℓ by making x sufficiently close to c. Notice that the definition puts the condition $0 < |x - c|$ on x, so that x is *not* allowed to take the value c. In other words we do not look at $x = c$, but rather at x *near* to c.

Also observe that we only consider the limit of f at a point c that is not isolated. In the exercises you will be asked to discuss why it would be nonsensical to use the above definition to study the limit at an isolated point.

EXAMPLE 5.3 Let $E = \mathbb{R} \setminus \{0\}$ and

$$f(x) = x \cdot \sin(1/x) \quad \text{if } x \in E.$$

See Figure 5.1. Then $\lim_{x \to 0} f(x) = 0$. To see this, let $\epsilon > 0$. Choose $\delta = \epsilon$. If $0 < |x - 0| < \delta$ then

$$|f(x) - 0| = |x \cdot \sin(1/x)| \leq |x| < \delta = \epsilon,$$

as desired. Thus the limit exists and equals 0.   ∎

EXAMPLE 5.4 Let $E = \mathbb{R}$ and

$$g(x) = \begin{cases} 1 & \text{if} \quad x \text{ is rational} \\ 0 & \text{if} \quad x \text{ is irrational.} \end{cases}$$

Then $\lim_{x \to c} g(x)$ does not exist for any point c of E.

To see this, fix $c \in \mathbb{R}$. Seeking a contradiction, assume that there is a limiting value ℓ for g at c. If this is so then we take $\epsilon = 1/2$ and we can find a $\delta > 0$ such that $0 < |x - c| < \delta$ implies

$$|g(x) - \ell| < \epsilon = \frac{1}{2}. \qquad (5.4.1)$$

If we take x to be rational then (5.4.1) says that

$$|1 - \ell| < \frac{1}{2}, \qquad (5.4.2)$$

while if we take x irrational then (5.4.1) says that

$$|0 - \ell| < \frac{1}{2}. \qquad (5.4.3)$$

But then the triangle inequality gives that

$$\begin{aligned} |1 - 0| &= |(1 - \ell) + (\ell - 0)| \\ &\leq |1 - \ell| + |\ell - 0|, \end{aligned}$$

which by (5.4.2) and (5.4.3) is

$$< 1.$$

This contradiction, that $1 < 1$, allows us to conclude that the limit does not exist at c. ∎

Proposition 5.5 *Let f be a function whose domain contains adjoining intervals (a, c) and (c, b). If $\lim_{x \to c} f(x) = \ell$ and $\lim_{x \to c} f(x) = m$, then $\ell = m$.*

Proof: Let $\epsilon > 0$. Choose $\delta_1 > 0$ such that, if $x \in E$ and $0 < |x - c| < \delta_1$, then $|f(x) - \ell| < \epsilon/2$. Similarly choose $\delta_2 > 0$ such that, if $x \in E$ and $0 < |x - c| < \delta_2$, then $|f(x) - m| < \epsilon/2$. Define δ to be the minimum of δ_1 and δ_2. If $x \in E$ and $0 < |x - c| < \delta$, then the triangle inequality tells us that

$$\begin{aligned} |\ell - m| &= |(\ell - f(x)) + (f(x) - m)| \\ &\leq |(\ell - f(x)| + |f(x) - m)| \\ &< \frac{\epsilon}{2} + \frac{\epsilon}{2} \\ &= \epsilon \end{aligned}$$

Since $|\ell - m| < \epsilon$ for every positive ϵ we conclude that $\ell = m$. That is the desired result. □

POINT OF CONFUSION 5.6 The point of the last proposition is that, if a limit is calculated by two different methods, then the same answer will result. While of primarily philosophical interest now, this will be important information later when we establish the existence of certain limits.

This is a good time to observe that the limits

$$\lim_{x \to c} f(x)$$

and

$$\lim_{h \to 0} f(c + h)$$

are equal in the sense that, if one limit exists then so does the other, and they both have the same value.

In order to facilitate checking that certain limits exist, we now record some elementary properties of the limit. This requires that we first recall how functions are combined.

Suppose that f and g are each functions which have domain E. We define the *sum* or *difference* of f and g to be the function

$$(f \pm g)(x) = f(x) \pm g(x) \,,$$

the *product* of f and g to be the function

$$(f \cdot g)(x) = f(x) \cdot g(x) \,,$$

and the quotient of f and g to be

$$\left(\frac{f}{g} \right)(x) = \frac{f(x)}{g(x)} \,.$$

Notice that the quotient is only defined at points x for which $g(x) \neq 0$. Now we have:

Theorem 5.7 (Elementary Properties of Limits of Functions) *Let f and g be functions whose domains contain adjoining intervals (a, c) and (c, b). Assume that*

(i) $\lim_{x \to c} f(x) = \ell$

(ii) $\lim_{x \to c} g(x) = m \,.$

Then

(a) $\lim_{x \to c} (f \pm g)(x) = \ell \pm m$

(b) $\lim\limits_{x \to c} (f \cdot g)(x) = \ell \cdot m$

(c) $\lim\limits_{x \to c} (f/g)(x) = \ell/m$ provided $m \neq 0$.

Proof: We prove part **(b)**. Parts **(a)** and **(c)** are treated in the exercises.

Let $\epsilon > 0$. We may also assume that $\epsilon < 1$. Choose $\delta_1 > 0$ such that, if $x \in E$ and $0 < |x - c| < \delta_1$, then

$$|f(x) - \ell| < \frac{\epsilon}{2(|m| + 1)}.$$

Choose $\delta_2 > 0$ such that, if $x \in E$ and $0 < |x - c| < \delta_2$ then

$$|g(x) - m| < \frac{\epsilon}{2(|\ell| + 1)}.$$

(Notice that this last inequality implies that $|g(x)| < |m| + |\epsilon|$.) Let δ be the minimum of δ_1 and δ_2. If $x \in E$ and $0 < |x - c| < \delta$, then

$$\begin{aligned}
|f(x) \cdot g(x) - \ell \cdot m| &= |(f(x) - \ell) \cdot g(x) + (g(x) - m) \cdot \ell| \\
&\leq |(f(x) - \ell) \cdot g(x)| + |(g(x) - m) \cdot \ell| \\
&< \left(\frac{\epsilon}{2(|m| + 1)} \right) \cdot |g(x)| + \left(\frac{\epsilon}{2(|\ell| + 1)} \right) \cdot |\ell| \\
&\leq \left(\frac{\epsilon}{2(|m| + 1)} \right) \cdot (|m| + |\epsilon|) + \frac{\epsilon}{2} \\
&< \frac{\epsilon}{2} + \frac{\epsilon}{2} \\
&= \epsilon.
\end{aligned}$$

$\square$

EXAMPLE 5.8 It is a simple matter to check that, if $f(x) = x$, then

$$\lim_{x \to c} f(x) = c$$

for every real c. (Indeed, for $\epsilon > 0$ we may take $\delta = \epsilon$.) Also, if $g(x) \equiv \alpha$ is the constant function taking value α, then

$$\lim_{x \to c} g(x) = \alpha.$$

It then follows from parts **(a)** and **(b)** of the theorem that, if $f(x)$ is any polynomial function, then

$$\lim_{x \to c} f(x) = f(c).$$

Moreover, if $r(x)$ is any *rational function* (quotient of polynomials) then we may also use part **(c)** of the theorem to conclude that

$$\lim_{x \to c} r(x) = r(c)$$

for all points c at which the rational function $r(x)$ is defined. ∎

EXAMPLE 5.9 If x is a small, positive real number then $0 < \sin x < x$. This is true because $\sin x$ is the nearest distance from the point $(\cos x, \sin x)$ to the x-axis while x is the distance from that point to the x-axis along an arc. If $\epsilon > 0$, then we set $\delta = \epsilon$. We conclude that, if $0 < |x - 0| < \delta$, then

$$|\sin x - 0| < |x| < \delta = \epsilon \,.$$

Since $\sin(-x) = -\sin x$, the same result holds when x is a negative number with small absolute value. Therefore

$$\lim_{x \to 0} \sin x = 0 \,.$$

Notice that
$$1 - \cos x = 2 \sin^2(x/2) \,.$$

Certainly $\lim_{x \to 0} \sin(x/2) = 0$ (exercise). So we may apply Theorem 5.7 to conclude that $\lim_{x \to 0} \cos x = 1$.

Now fix any real number c. We have

$$\begin{aligned}
\lim_{x \to c} \sin x &= \lim_{h \to 0} \sin(c + h) \\
&= \lim_{h \to 0} \left(\sin c \cos h + \cos c \sin h \right) \\
&= \sin c \cdot 1 + \cos c \cdot 0 \\
&= \sin c \,.
\end{aligned}$$

We of course have used parts **(a)** and **(b)** of the theorem to commute the limit process with addition and multiplication. A similar argument shows that

$$\lim_{x \to c} \cos x = \cos c \,.$$

We conclude that sine and cosine are continuous functions. ∎

Remark 5.10 In the last example, we have used the definition of the sine function and the cosine function that you learned in calculus. In Chapter 9, when we learn about series of functions, we will learn a more rigorous method for treating the trigonometric functions.

We conclude by giving a characterization of the limit of a function using sequences.

Proposition 5.11 *Let f be a function whose domain E contains adjoining intervals (a, c) and (c, b). Then*

$$\lim_{x \to c} f(x) = \ell \tag{5.11.1}$$

if and only if, for any sequence $\{a_j\}$ satisfying $\lim_{j \to \infty} a_j = c$, it holds that

$$\lim_{j \to \infty} f(a_j) = \ell. \tag{5.11.2}$$

Proof: Assume that condition (5.11.1) fails. Then there is an $\epsilon > 0$ such that for no $\delta > 0$ is it the case that when $0 < |x - c| < \delta$ then $|f(x) - \ell| < \epsilon$. Thus, for each $\delta = 1/j$, we may choose a number $a_j \in E \setminus \{c\}$ with $0 < |a_j - c| < 1/j$ and $|f(a_j) - \ell| \geq \epsilon$. But then condition (5.11.2) fails for this sequence $\{a_j\}$.

If condition (5.11.2) fails then there is a sequence $\{a_j\}$ such that $\lim_{j \to \infty} a_j = c$ but $\lim_{j \to \infty} f(a_j) \neq \ell$. This means that there is an $\epsilon > 0$ such that, for infinitely many a_j, it holds that $|f(a_j) - \ell| \geq \epsilon$. But then, no matter how small $\delta > 0$, there will be an a_j satisfying $0 < |a_j - c| < \delta$ (since $a_j \to c$) and $|f(a_j) - \ell| \geq \epsilon$. Thus (5.11.1) fails. $\qquad\square$

A Look Back

1. Give a verbal description of what a limit is.

2. Give an example of a function on $\mathbb{R}$ that does not have a limit at any point.

3. Give an example of a function on $\mathbb{R}$ that has a limit at every point except the origin.

4. Explain how the rigorous definition of limit using ϵ and δ relates to the intuitive definition of limit that uses the phrase "can draw the graph without lifting your pencil from the paper."

Exercises

1. Let f and g be functions on a set $A = (a, c) \cup (c, b)$ and assume that $f(x) \leq g(x)$ for all $x \in A$. Assuming that both limits exist, show that

$$\lim_{x \to c} f(x) \leq \lim_{x \to c} g(x).$$

Does the conclusion improve if we assume that $f(x) < g(x)$ for all $x \in A$?

2. Prove parts (a) and (c) of Theorem 5.7.

3. Show that the limit of a function f at a point c is unique.

4. Give a definition of limit using the concept of distance.

4. If $\lim_{x \to c} f(x) = \ell > 0$ then prove that there is a $\delta > 0$ so small that $|x - c| < \delta$ guarantees that $f(x) > \ell/2$.

5. Show that, if f is a monotone function, then f has a limit at "most" points. What does the word "most" mean in this context?

6. Give an example of a function such that $\lim_{x \to c} f(x)$ exists at every point but f is discontinuous at every point.

7. Give an example of a function that is discontinuous at every point of its domain.

8. Prove that $\lim_{x \to c} f(x) = \lim_{h \to 0} f(c + h)$ whenever both expressions make sense.

9. Show that the set of jump discontinuities of a function $f : \mathbb{R} \to \mathbb{R}$ is at most countable.

* **10.** Give a definition of limit using the concept of open set.

* **11.** Give an example of a function $f : \mathbb{R} \to \mathbb{R}$ so that $\lim_{x \to c} f(x)$ exists when c is irrational but does not exist when c is rational.

5.2 Continuous Functions

Preliminary Remarks

The concept of continuous function is intuitively appealing. But what we need is a rigorous definition. Having the idea of limit under control enables us to give a precise and accurate definition of continuity. Then we can prove some results about continuous functions and begin to develop a cogent theory.

Definition 5.12 Let $E \subseteq \mathbb{R}$ be a set and let f be a real-valued function with domain E. Fix a point c which is in E. We say that f is *continuous* at c if

$$\lim_{x \to c} f(x) = f(c).$$

We learned from the penultimate example of Section 5.1 that polynomial functions are continuous at every real x. So are the transcendental functions $\sin x$ and $\cos x$ (see Example 5.9). A rational function is continuous at every point of its domain.

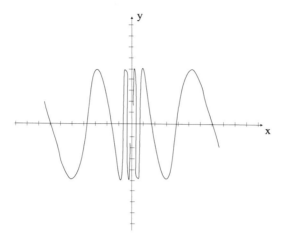

Figure 5.2: A function discontinuous at 0.

EXAMPLE 5.13 The function

$$h(x) = \begin{cases} \sin(1/x) & \text{if} \quad x \neq 0 \\ 1 & \text{if} \quad x = 0 \end{cases}$$

is discontinuous at 0. See Figure 5.2. The reason is that

$$\lim_{x \to 0} h(x)$$

does not exist. (Details of this assertion are left for you: notice that $h(1/(j\pi)) = 0$ while $h(2/[(4j + 1)\pi]) = 1$ for $j = 1, 2, \ldots$.)

The function

$$k(x) = \begin{cases} x \cdot \sin(1/x) & \text{if} \quad x \neq 0 \\ 1 & \text{if} \quad x = 0 \end{cases}$$

is also discontinuous at $x = 0$. This time the limit $\lim_{x \to 0} k(x)$ exists (see Example 5.3), but the limit does not agree with $k(0)$.

However, the function

$$m(x) = \begin{cases} x \cdot \sin(1/x) & \text{if} \quad x \neq 0 \\ 0 & \text{if} \quad x = 0 \end{cases}$$

is continuous at $x = 0$ because the limit at 0 exists and agrees with the value of the function there. See Figure 5.3. ∎

POINT OF CONFUSION 5.14 As we shall see in detail below, a function can be discontinuous because it oscillates, or it can be discontinuous because the limit at c disagrees with the value at c. We must learn to distinguish these two cases.

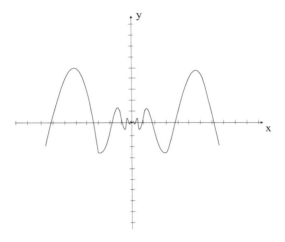

Figure 5.3: A function continuous at 0.

The arithmetic operations $+, -, \times$, and $\div$ preserve continuity (so long as we avoid division by zero). We now formulate this assertion as a theorem.

Theorem 5.15 *Let f and g be functions with domain E and let c be a point of E. If f and g are continuous at c then so are $f \pm g, f \cdot g$, and (provided $g(c) \neq 0$) f/g.*

Proof: Apply Theorem 5.7 of Section 5.1.   □

Continuous functions may also be characterized using sequences:

Proposition 5.16 *Let f be a function with domain E and fix $c \in E$. The function f is continuous at c if and only if, for every sequence $\{a_j\} \subseteq E$ satisfying $\lim_{j \to \infty} a_j = c$, it holds that*

$$\lim_{j \to \infty} f(a_j) = f(c).$$

Proof: Apply Proposition 5.11 of Section 5.1.   □

POINT OF CONFUSION 5.17 A continuous function is one with the property that what it does at a point c is what we anticipate that it will do at that point c. What does this mean?

What we anticipate that the function will do is the *limit* as $x \to c$. What it actually does at the point is take the value $f(c)$. We demand for continuity that the two agree.

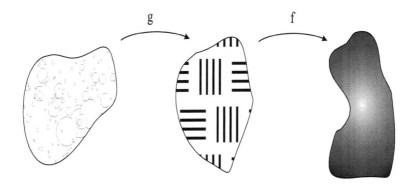

Figure 5.4: Composition of functions.

Recall that, if g is a function with domain D and range E, and if f is a function with domain E and range F, then the *composition* of f and g is

$$f \circ g(x) = f(g(x)).$$

See Figure 5.4.

Proposition 5.18 *Let g have domain D and range E and let f have domain E and range F. Let $c \in D$. Assume that g is continuous at c and that f is continuous at $g(c)$. Then $f \circ g$ is continuous at c.*

Proof: Let $\{a_j\}$ be any sequence in D such that $\lim_{j \to \infty} a_j = c$. Then

$$\lim_{j \to \infty} f \circ g(a_j) = \lim_{j \to \infty} f(g(a_j)) = f\left(\lim_{j \to \infty} g(a_j)\right)$$

$$= f\left(g\left(\lim_{j \to \infty} a_j\right)\right) = f(g(c)) = f \circ g(c).$$

Now apply Proposition 5.11. □

POINT OF CONFUSION 5.19 Continuity is a robust attribute for a function. It is preserved under most of our standard operations. Limits are a bit more delicate, and must be handled more carefully. See the next remark.

Remark 5.20 It is not the case that if

$$\lim_{x \to c} g(x) = \ell$$

and

$$\lim_{t \to \ell} f(t) = m$$

then

$$\lim_{x \to c} f \circ g(x) = m \, .$$

A counterexample is given by the functions

$$g(x) = 0$$

$$f(x) = \begin{cases} 2 & \text{if} \quad x \neq 0 \\ 5 & \text{if} \quad x = 0. \end{cases}$$

Notice that $\lim_{x \to 0} g(x) = 0$, $\lim_{t \to 0} f(t) = 2$, yet $\lim_{x \to 0} f \circ g(x) = 5$.

The additional hypothesis that f be continuous at ℓ is necessary in order to guarantee that the limit of the composition will behave as expected.

A Look Back

1. Describe in words what a continuous function is.

2. Describe what the role of limits is in the definition of continuity.

3. Why do we need to use the concept of limit to define continuity?

4. Why is it the case that a continuous function is not necessarily differentiable?

Exercises

1. Let $0 < \alpha \leq 1$. A function f with domain E is said to satisfy a *Lipschitz condition* of order α if there is a constant $C > 0$ such that, for any $s, t \in E$, it holds that $|f(s) - f(t)| \leq C \cdot |s - t|^\alpha$. Prove that such a function must be uniformly continuous (see the next section for the definition of this concept).

2. Define the function

$$g(x) = \begin{cases} 0 & \text{if} \quad x \text{ is irrational} \\ x & \text{if} \quad x \text{ is rational} \end{cases}$$

At which points x is g continuous? At which points is it discontinuous?

3. Explain why it would be foolish to define the concept of limit at an isolated point.

4. Using examples, explain why the continuous image of a connected set is connected.

5. Explain why a continuous function defined on a compact set must be bounded.

6. Explain why a continuous function defined on a compact set actually assumes its maximum and minimum values.

7. Give an example of a closed set E and a continuous function f so that $f(E)$ is open.

8. Give an example of an open set U and a continuous function f so that $f(U)$ is closed.

* 9. Let f be a continuous function whose domain contains an open interval (a, b). What form can $f((a, b))$ have? (**Hint:** There are just four possibilities.)

* 10. Let f be a continuous function on the open interval (a, b). Under what circumstances can f be extended to a continuous function on $[a, b]$?

* 11. Define continuity using the notion of closed set.

* 12. The image of a compact set under a continuous function is compact (see the next section). But the image of a closed set need not be closed. Explain. The *inverse image* of a compact set under a continuous function need not be compact. Explain.

5.3 Topological Properties and Continuity

Preliminary Remarks

Continuous functions are, from the point of view of topology, very natural objects to study. Continuous functions interact very naturally with open sets, with compact sets, and with other standard artifacts of this theory. We explore these connections in the present section.

Definition 5.21 Let f be a function with domain E and let L be a subset of E. We define

$$f(L) = \{f(x) : x \in L\}.$$

The set $f(L)$ is called the *image* of L under f. See Figure 5.5.

Theorem 5.22 *The image of a compact set under a continuous function is also compact.*

Proof: Let K be a compact set and f a continuous function. Consider the set $K' \equiv f(K)$. Let $\{x_j\}$ be a sequence in K'. Then each x_j has a pre-image t_j (that is to say, $f(t_j) = x_j$) in K. Since K is compact, there is a subsequence t_{j_k} that converges to a point $t_0 \in K$. But then, by continuity, the points $f(t_{j_k})$ converge to $f(t_0) \in K'$. So we see that $\{x_j\}$ has a convergent subsequence $x_{j_k} = f(t_{j_k})$. Hence K' is compact. □

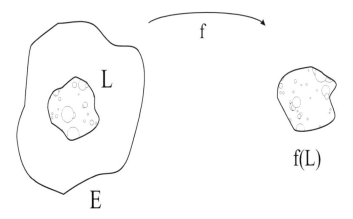

Figure 5.5: The image of the set L under the function f.

POINT OF CONFUSION 5.23 It is not the case that the continuous image of a closed set is closed. For instance, take $f(x) = 1/(1 + x^2)$ and $E = \mathbb{R}$: the set E is closed and f is continuous but $f(E) = (0, 1]$ is not closed.

It is also not the case that the continuous image of a bounded set is bounded. As an example, take $f(x) = 1/x$ and $E = (0, 1)$. Then E is bounded and f continuous but $f(E) = (1, \infty)$ is unbounded.

However, the combined properties of closedness *and* boundedness (that is, compactness) are preserved. That is the content of the preceding theorem.

Corollary 5.24 *Let f be a continuous, real-valued function with compact domain $\subseteq \mathbb{R}$. Then there is a number L such that*

$$|f(x)| \leq L$$

for all $x \in K$.

Proof: We know from the theorem that $f(K)$ is compact. By Theorem 4.30, we conclude that $f(K)$ is bounded. Thus there is a number L such that $|t| \leq L$ for all $t \in f(K)$. But that is just the assertion that we wish to prove. $\square$

In fact we can prove an important strengthening of the corollary. Since $f(K)$ is compact, it contains its supremum M and its infimum m. Therefore there must be a number $C \in K$ such that $f(C) = M$ and a number $c \in K$ such that $f(c) = m$. In other words, $f(c) \leq f(x) \leq f(C)$ for all $x \in K$. We summarize:

Theorem 5.25 *Let f be a continuous function on a compact set $K \subseteq \mathbb{R}$. Then there exist numbers c and C in K such that $f(c) \leq f(x) \leq f(C)$ for* all $x \in K$. We call c an *absolute minimum* for f on K and C an *absolute maximum* for f on K. We call $f(c)$ the *absolute minimum value* for f on K and $f(C)$ the *absolute maximum value* for f on K.

Notice that, in the last theorem, the location of the absolute maximum and absolute minimum need not be unique. For instance, the function $\sin x$ on the compact interval $[0, 4\pi]$ has an absolute minimum at $3\pi/2$ and $7\pi/2$. It has an absolute maximum at $\pi/2$ and at $5\pi/2$.

Now we define a refined type of continuity called "uniform continuity." We shall learn that this new notion of continuous function arises naturally for a continuous function on a compact set. It will also play an important role in our later studies, especially in the context of the integral.

Definition 5.26 Let f be a function with domain $E \subseteq \mathbb{R}$. We say that f is *uniformly continuous* on E if, for each $\epsilon > 0$, there is a $\delta > 0$ such that, whenever $s, t \in E$ and $|s - t| < \delta$, then $|f(s) - f(t)| < \epsilon$.

POINT OF CONFUSION 5.27 Observe that "uniform continuity" differs from "continuity" in that it treats all points of the domain simultaneously: the $\delta > 0$ that is chosen is independent of the points $s, t \in E$. This difference is highlighted by the next two examples.

EXAMPLE 5.28 Suppose that a function $f : \mathbb{R} \to \mathbb{R}$ satisfies the condition

$$|f(s) - f(t)| \leq C \cdot |s - t|, \tag{5.28.1}$$

where C is some positive constant. This is called a *Lipschitz condition*, and it arises frequently in analysis. We refer to the collection of functions satisfying such a condition as the space Lip_1. Let $\epsilon > 0$ and set $\delta = \epsilon/C$. If $|x - y| < \delta$ then, by (5.28.1),

$$|f(x) - f(y)| \leq C \cdot |x - y| < C \cdot \delta = C \cdot \frac{\epsilon}{C} = \epsilon.$$

It follows that f is uniformly continuous. ∎

EXAMPLE 5.29 Consider the function $f(x) = x^2$. Fix a point $c \in \mathbb{R}, c > 0$, and let $\epsilon > 0$. In order to guarantee that $|f(x) - f(c)| < \epsilon$ we must have (for $x > 0$)

$$|x^2 - c^2| < \epsilon$$

or

$$|x - c| < \frac{\epsilon}{x + c}.$$

Since x will range over a neighborhood of c, we see that the required δ in the definition of continuity cannot be larger than $\epsilon/(2c)$. In fact the choice $|x - c| < \delta = \epsilon/(2c + 1)$ will do the job. We see that the choice of δ depends decisively on c.

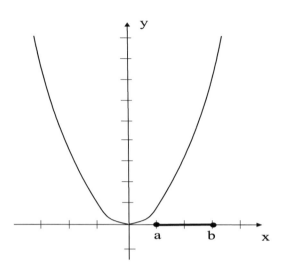

Figure 5.6: Uniform continuity on the interval $[a, b]$.

Put in slightly different words, let $\epsilon = 1$. Then

$$|f(j + 1/j) - f(j)| \geq |(j + 1/j)^2 - j^2| > 1 = \epsilon$$

for any j. Thus, for this ϵ, we may not take δ to be $1/j$ for any j. So no uniform δ exists.

Thus the choice of δ depends not only on ϵ (which we have come to expect) but also on c. In particular, f is not uniformly continuous on $\mathbb{R}$. This a quantitative reflection of the fact that the graph of f becomes ever steeper as the variable moves to the right.

Notice that the same calculation shows that the function f with restricted domain $[a, b], 0 < a < b < \infty$, is uniformly continuous. That is because, when the function is restricted to $[a, b]$, its rate of increase (sometimes called the "slope") does not become arbitrarily large. See Figure 5.6.   ∎

Now the main result about uniform continuity is the following:

Theorem 5.30 *Let f be a continuous function with compact domain K. Then f is uniformly continuous on K.*

Proof: Suppose not. Then there is an $\epsilon > 0$ and points x_j, t_j such that $|x_j - t_j| < 1/j$ yet $|f(x_j) - f(t_j)| > \epsilon$. Since K is compact, there are subsequences x_{j_k} and t_{j_k} that converge respectively to points x_0 and t_0 in K. But in fact x_0 and t_0 must be the same point, yet $f(x_0) \neq f(t_0)$. That is a contradiction.   □

EXAMPLE 5.31 The function $f(x) = \sin(1/x)$ is continuous on the domain $E = (0, \infty)$ since it is the composition of continuous functions (refer again to Figure 5.2). However, it is not uniformly continuous since

$$\left| f\left(\frac{1}{2j\pi}\right) - f\left(\frac{1}{\frac{(4j+1)\pi}{2}}\right) \right| = 1$$

for $j = 1, 2, \ldots$. Thus, even though the arguments are becoming arbitrarily close together, the images of these arguments remain bounded apart. We conclude that f cannot be uniformly continuous. See Figure 5.2.

However, if f is considered as a function on any interval of the form $[a, b], 0 < a < b < \infty$, then the preceding theorem tells us that the function f is uniformly continuous. ∎

As an exercise, you should check that

$$g(x) = \begin{cases} x \sin(1/x) & \text{if } x \neq 0 \\ 0 & \text{if } x = 0 \end{cases}$$

is uniformly continuous on any interval of the form $[-N, N]$. See Figure 5.3.

Let us note that a function f is said to be continuous on a closed interval $[a, b]$ if f is continuous at each point of $[a, b]$. Refer back to our original definition of continuity.

Corollary 5.32 (The Intermediate Value Theorem) *Let f be a continuous function whose domain contains the interval $[a, b]$. Let γ be a number that lies between $f(a)$ and $f(b)$. Then there is a number c between a and b such that $f(c) = \gamma$. Refer to Figure 5.7.*

Proof: We merely sketch the proof.

Assume without loss of generality that $f(a) < 0$ and $f(b) > 0$ and $\gamma = 0$. Let

$$S = \{x \in [a, b] : f(x) < 0\}.$$

Then S is bounded and nonempty (because $a \in S$), so S has a least upper bound c. We claim that $f(c) = 0$.

Clearly $f(c) \leq 0$ by the continuity of f. If in fact $f(c) < 0$, then points x to the right of c will satisfy $f(x) < 0$, contradicting the fact that c was chosen to be the least upper bound of S. So it must be that $f(c) = 0$ as desired. □

Remark 5.33 Another way to think about the Intermediate Value Theorem is in terms of connectedness. The interval $[a, b]$ is of course connected. So its image $f([a, b])$ is connected. Therefore $f([a, b])$ must be an interval. Since that interval contains the points $f(a)$ and $f(b)$, it must therefore also contain the point γ which lies between $f(a)$ and $f(b)$.

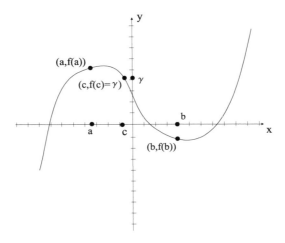

Figure 5.7: The Intermediate Value Theorem.

A Look Back

1. Say in words why the continuous image of a compact set is compact.

2. Say in words why the continuous image of a connected set is connected.

3. Say in words why the continuous image of a bounded set is not necessarily bounded.

4. Say in words why the continuous image of a closed set is not necessarily closed.

Exercises

1. If f is continuous on $[0,1]$ and if $f(x)$ is positive for each rational x, then does it follow that f is positive at all x?

2. Is the inverse image of a connected set connected?

3. Let S be any subset of $\mathbb{R}$. Define the function

$$f(x) = \inf\{|x - s| : s \in S\}.$$

[We think of $f(x)$ as the distance of x to S.] Prove that f is uniformly continuous.

3. Define the function $g(x)$ to take the value 0 at irrational values of x and to take the value $1/q$ when $x = p/q$ is a rational number in lowest terms, $q > 0$. At which points is g continuous? At which points is the function discontinuous?

4. Let f be any function whose domain and range is the entire real line. If A and B are disjoint sets does it follow that $f(A)$ and $f(B)$ are disjoint sets? If C and D are disjoint sets does it follow that $f^{-1}(C)$ and $f^{-1}(D)$ are disjoint?

5. Let f be any function whose domain is the entire real line. If A and B are sets then is $f(A \cup B) = f(A) \cup f(B)$? If C and D are sets then is $f^{-1}(C \cup D) = f^{-1}(C) \cup f^{-1}(D)$? What is the answer to these questions if we replace $\cup$ by $\cap$?

6. Prove that the function $f(x) = \sin x$ can be written, on the interval $(0, 2\pi)$, as the difference of two increasing functions.

7. Let f be a continuous function with domain $[0, 1]$ and range $[0, 1]$. Prove that there exists a point $c \in [0, 1]$ such that $f(c) = c$. (**Hint:** Apply the Intermediate Value Theorem to the function $g(x) = f(x) - x$.) Prove that this result is false if the domain and range of the function are both $(0, 1)$.

8. Let f be a continuous function and let $\{a_j\}$ be a Cauchy sequence in the domain of f. Does it follow that $\{f(a_j)\}$ is a Cauchy sequence? What if we assume instead that f is uniformly continuous?

9. Let E and F be disjoint closed sets of real numbers. Prove that there is a continuous function f with domain the real numbers such that $\{x : f(x) = 0\} = E$ and $\{x : f(x) = 1\} = F$.

10. If K and L are sets then define

$$K + L = \{k + \ell : k \in K \text{ and } \ell \in L\}.$$

If K and L are compact then prove that $K + L$ is compact. If K and L are merely closed, does it follow that $K + L$ is closed?

11. Let A be a finite disjoint union of m intervals. Let f be a continuous function. Then what can you say about $f(A)$?

* 11. Let E be any closed set of real numbers. Prove that there is a continuous function f with domain $\mathbb{R}$ such that $\{x : f(x) = 0\} = E$.

* 12. Give an example of a continuous function f and a connected set E such that $f^{-1}(E)$ is not connected. Is there a condition you can add that will force $f^{-1}(E)$ to be connected?

* 13. Give an example of a continuous function f and an open set U so that $f(U)$ is not open.

14. Let E be a closed set in $\mathbb{R}$. Define a continuous function f that is equal to 0 on E and is positive on the complement of E.

* 15. A function f with domain A and range B is called a *homeomorphism* if it is one-to-one, onto, continuous, and has continuous inverse. If such an f exists then we say that A and B are *homeomorphic*. Which sets of reals are homeomorphic to the open unit interval $(0, 1)$? Which sets of reals are homeomorphic to the closed unit interval $[0, 1]$?

5.4   Classifying Discontinuities and Monotonicity

Preliminary Remarks

Just by using a little logic, we can classify the discontinuities of a real function. This gives an elegant way to understand discontinuities. We also learn a nice theorem of Darboux about discontinuities of the derivative of a function.

We begin by refining our notion of limit:

Definition 5.34 Fix $c \in \mathbb{R}$. Let f be a function whose domain contains an interval (a, c). We say that f has *left limit* ℓ at c, and write

$$\lim_{x \to c^-} f(x) = \ell \,,$$

if, for every $\epsilon > 0$, there is a $\delta > 0$ such that, whenever $c - \delta < x < c$, then it holds that

$$|f(x) - \ell| < \epsilon \,.$$

Now suppose instead that the domain of f contains an interval (c, b). We say that f has *right limit* m at c, and write

$$\lim_{x \to c^+} f(x) = m$$

if, for every $\epsilon > 0$, there is a $\delta > 0$ such that, whenever $c < x < c + \delta$, then it holds that

$$|f(x) - m| < \epsilon \,.$$

This definition simply formalizes the notion of either letting x tend to c from the left only or from the right only.

Definition 5.35 Fix $c \in \mathbb{R}$. Let f be a function with domain E. Suppose that c is a limit point of $E \cap [c - \delta, c)$ for some $\delta > 0$ and that c is an element of E. We say that f is *left continuous* at c if

$$\lim_{x \to c^-} f(x) = f(c) \,.$$

Likewise, in case c is a limit point of $E \cap (c, c + \delta]$ for some $\delta > 0$ and is also an element of E, we say that f is *right continuous* at c if

$$\lim_{x \to c^+} f(x) = f(c) \,.$$

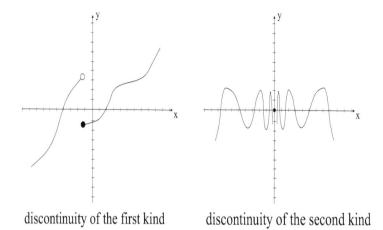

discontinuity of the first kind   discontinuity of the second kind

Figure 5.8: Discontinuities of the first and second kind.

Let f be a function with domain E. Let c in E and assume that f is discontinuous at c. There are two ways in which this discontinuity can occur:

I. If $\lim_{x \to c^-} f(x)$ and $\lim_{x \to c^+} f(x)$ both exist but either do not equal each other or do not equal $f(c)$ then we say that f has a *discontinuity of the first kind* (or sometimes a *simple discontinuity*) at c.

II. If either $\lim_{x \to c^-}$ does not exist or $\lim_{x \to c^+}$ does not exist then we say that f has a *discontinuity of the second kind* at c.

Refer to Figure 5.8.

EXAMPLE 5.36 Define

$$f(x) = \begin{cases} \sin(1/x) & \text{if} \quad x \neq 0 \\ 0 & \text{if} \quad x = 0 \end{cases}$$

$$g(x) = \begin{cases} 1 & \text{if} \quad x > 0 \\ 0 & \text{if} \quad x = 0 \\ -1 & \text{if} \quad x < 0 \end{cases}$$

$$h(x) = \begin{cases} 1 & \text{if } x \text{ is irrational} \\ 0 & \text{if } x \text{ is rational} \end{cases}$$

Then f has a discontinuity of the second kind at 0 while g has a discontinuity of the first kind at 0. The function h has a discontinuity of the second kind at every point. ∎

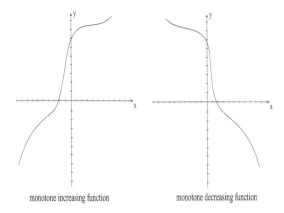

monotone increasing function   monotone decreasing function

Figure 5.9: Increasing and decreasing functions.

POINT OF CONFUSION 5.37 For a bounded function, a discontinuity of the second kind can be thought of as an oscillatory discontinuity. A discontinuity of the first kind can be thought of as a case of the limiting value not equaling the actual value (sometimes we think of this as a jump discontinuity).

Definition 5.38 Let f be a function whose domain contains an open interval (a, b). We say that f is *increasing* on (a, b) if, whenever $a < s < t < b$, it holds that $f(s) \leq f(t)$. We say that f is *decreasing* on (a, b) if, whenever $a < s < t < b$, it holds that $f(s) \geq f(t)$. See Figure 5.9.

If a function is either increasing or decreasing then we call it *monotone* or *monotonic*. Compare with the definition of monotonic sequences in Section 2.1.

As with sequences, the word "monotonic" is superfluous in many contexts. But its use is traditional and occasionally convenient.

Proposition 5.39 *Let f be a monotonic function on an open interval (a, b). Then all of the discontinuities of f are of the first kind.*

Proof: It is enough to show that, for each $c \in (a, b)$, the limits

$$\lim_{x \to c^-} f(x)$$

and

$$\lim_{x \to c^+} f(x)$$

exist.

Let us first assume that f is monotonically increasing. Fix $c \in (a, b)$. If $a < s < c$ then $f(s) \leq f(c)$. Therefore $S = \{f(s) : a < s < c\}$ is bounded above. Let M be the least upper bound of S. Pick $\epsilon > 0$. By definition of least upper bound there must be an $f(s) \in S$ such that $|f(s) - M| < \epsilon$. Let $\delta = |c - s|$. If $c - \delta < t < c$ then $s < t < c$ and $f(s) \leq f(t) \leq M$ or $|f(t) - M| < \epsilon$. Thus $\lim_{x \to c^-} f(x)$ exists and equals M.

If we set m equal to the infimum of the set $T = \{f(t) : c < t < b\}$ then a similar argument shows that $\lim_{x \to c^+} f(x)$ exists and equals m.

So we see that the function f has both a left and a right limit at c. So either f is continuous at c or f has a discontinuity of the first kind.

The argument for f monotonically decreasing is the same, and we omit the details. $\qquad\square$

Corollary 5.40 *Let f be a monotonic function on an interval (a, b). Then f has at most countably many discontinuities.*

Proof: Assume for simplicity that f is monotonically increasing. If c is a discontinuity then the proposition tells us that

$$\lim_{x \to c^-} f(x) < \lim_{x \to c^+} f(x).$$

Therefore there is a rational number q_c between $\lim_{x \to c^-} f(x)$ and $\lim_{x \to c^+} f(x)$. Notice that different discontinuities will have different rational numbers associated to them because if $\hat{c}$ is another discontinuity and, say, $\hat{c} < c$ then

$$\lim_{x \to \hat{c}^-} f(x) < q_{\hat{c}} < \lim_{x \to \hat{c}^+} f(x) \leq \lim_{x \to c^-} f(x) < q_c < \lim_{x \to c^+} f(x).$$

Thus we have exhibited a one-to-one function of the set of discontinuities of f into the set of rational numbers. It follows (see Appendix II) that the set of discontinuities is countable. $\qquad\square$

POINT OF CONFUSION 5.41 A function can have arbitrarily many (even uncountably many) discontinuities. But a monotone function has at most countably many.

Theorem 5.42 *Let f be a strictly monotone, continuous function with domain $[a, b]$. Then f^{-1} exists and is continuous.*

Proof: Assume without loss of generality that f is strictly monotone *increasing*. Let us extend f to the entire real line by defining

$$f(x) = \begin{cases} (x - a) + f(a) & \text{if} \quad x < a \\ \text{as given} & \text{if} \quad a \leq x \leq b \\ (x - b) + f(b) & \text{if} \quad x > b. \end{cases}$$

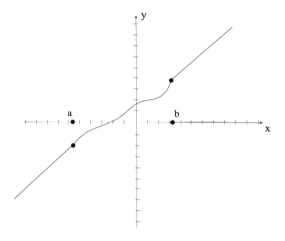

Figure 5.10: A strictly monotonically increasing function.

See Figure 5.10. Then it is easy to see that this extended version of f is still continuous and is strictly monotone increasing on all of $\mathbb{R}$.

The definition of strict monotonicity implies certainly that f is one-to-one. That it is onto follows from the continuity and from the way that we extended the function to have domain all of $\mathbb{R}$. Thus f^{-1} exists.

The extended function f takes any open interval (c, d) to the open interval $(f(c), f(d))$. Since any open set is a union of open intervals, we see that f takes any open set to an open set. In other words, $\left[f^{-1}\right]^{-1}$ takes open sets to open sets. But this just says that f^{-1} is continuous.

Since the inverse of the extended function f is continuous, then so is the inverse of the original function f. That completes the proof. □

A Look Back

1. What is a discontinuity of the first kind?

2. What is a discontinuity of the second kind?

3. Why can a discontinuity not be both of the first kind and the second kind?

4. Why is there no other kind of discontinuity besides first and second?

Exercises

1. Give an example of two functions, discontinuous at $x = 0$, whose sum *is* continuous at $x = 0$. Give an example of two such functions whose product is

continuous at $x = 0$. How does the problem change if we replace "product" by "quotient"?

2. Let f be a function with domain $\mathbb{R}$. If $f^2(x) = f(x) \cdot f(x)$ is continuous, then does it follow that f is continuous? If $f^3(x) = f(x) \cdot f(x) \cdot f(x)$ is continuous, then does it follow that f is continuous? What about if both f^2 and f^3 are continuous?

3. Fix an interval (a, b). Is the collection of increasing functions on (a, b) closed under $+, -, \times$, or $\div$?

4. Let f be a continuous function whose domain contains a closed, bounded interval $[a, b]$. What topological properties does $f([a, b])$ possess? Is this set necessarily an interval?

5. Refer to Exercise **15** of Section 5.3 for terminology. Show that there is no homeomorphism from the real line to the interval $[0, 1)$.

6. Let f be a function with domain $\mathbb{R}$. Prove that the set of discontinuities of the first kind for f is countable. (**Hint:** If the left and right limits at a point disagree then you can slip a rational number between them.)

7. Let f be a function with domain $(-1, 1)$. Can the set of discontinuities of f of the first kind be countable? Uncountable? What about the set of discontinuities of the second kind?

8. Let f and g be functions and assume that each has a discontinuity of the first kind at the origin. What can you say about the behavior of $f + g$ at the origin?

* **9.** Let A be any left-to-right ordered, countable subset of the reals. Construct an increasing function whose set of points of discontinuity is precisely the set A. Explain why this is, in general, impossible for an uncountable set A.

* **10.** TRUE or FALSE: If f is a function with domain and range the real numbers and which is both one-to-one and onto, then f must be either increasing or decreasing. Does your answer change if we assume that f is continuously differentiable?

* **11.** Let $I \subseteq \mathbb{R}$ be an open interval and $f : I \to \mathbb{R}$ a function. We say that f is *convex* if whenever $\alpha, \beta \in I$ and $0 \le t \le 1$ then

$$f((1 - t)\alpha + t\beta) \le (1 - t)f(\alpha) + tf(\beta).$$

Prove that a convex function must be continuous. What does this definition of convex function have to do with the notion of "concave up" that you learned in calculus?

* **12.** Refer to Exercise **11** for terminoloogy. What can you say about differentiability of a convex function?

Chapter 6

Differentiation of Functions

6.1 The Concept of Derivative

Preliminary Remarks
Of course the derivative is a significant idea from calculus, and it is important that we establish a rigorous and precise understanding of the concept. Our aim is to develop the derivative as a precise analytic tool, and also to get a rigorous proof of the Fundamental Theorem of Calculus.

Let f be a function with domain an open interval I. If $x \in I$ then the quantity

$$\frac{f(t) - f(x)}{t - x}$$

measures the slope of the chord of the graph of f that connects the points $(x, f(x))$ and $(t, f(t))$. See Figure 6.1. If we let $t \to x$ then the limit of the quantity represented by this "Newton quotient" should represent the slope of the graph *at the point* x. These considerations motivate the definition of the derivative:

Definition 6.1 If f is a function with domain an open interval I and if $x \in I$, then the limit

$$\lim_{t \to x} \frac{f(t) - f(x)}{t - x},$$

when it exists, is called the *derivative* of f at x. See Figure 6.2. If the derivative of f at x exists then we say that f is *differentiable* at x. If f is differentiable at every $x \in I$ then we say that f is *differentiable on I*.

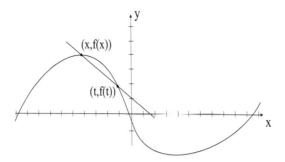

Figure 6.1: The Newton quotient.

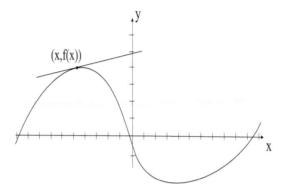

Figure 6.2: The derivative.

We write the derivative of f at x either as

$$f'(x) \quad \text{or} \quad \frac{d}{dx}f \quad \text{or} \quad \frac{df}{dx} \quad \text{or} \quad \dot{f}.$$

The last of these is particularly popular among physicists (as an homage to Isaac Newton).

We begin our discussion of the derivative by establishing some basic properties and relating the notion of derivative to continuity.

Lemma 6.2 *If f is differentiable at a point x then f is continuous at x. In particular, $\lim_{t \to x} f(t) = f(x)$.*

Proof: We use Theorem 5.7(**b**) about limits to see that

$$
\begin{aligned}
\lim_{t \to x} (f(t) - f(x)) &= \lim_{t \to x} \left((t - x) \cdot \frac{f(t) - f(x)}{t - x} \right) \\
&= \lim_{t \to x} (t - x) \cdot \lim_{t \to x} \frac{f(t) - f(x)}{t - x} \\
&= 0 \cdot f'(x) \\
&= 0.
\end{aligned}
$$

Therefore $\lim_{t \to x} f(t) = f(x)$ and f is continuous at x.   □

POINT OF CONFUSION 6.3 All differentiable functions are continuous: differentiability is a stronger property than continuity. Observe that the function $f(x) = |x|$ is continuous at every x but is not differentiable at 0. So continuity does not imply differentiability. Details appear in Example 6.6 below.

POINT OF CONFUSION 6.4 You might think, from your experience in other math courses, that most functions are differentiable at most points. As we shall learn below, this expectation is woefully incorrect. It turns out that "most" continuous functions are not differentiable at any point.

Theorem 6.5 *Assume that f and g are functions with domain an open interval I and that f and g are differentiable at $x \in I$. Then $f \pm g, f \cdot g$, and f/g are differentiable at x (for f/g we assume that $g(x) \neq 0$). Moreover*

(**a**) $(f \pm g)'(x) = f'(x) \pm g'(x);$

(**b**) $(f \cdot g)'(x) = f'(x) \cdot g(x) + f(x) \cdot g'(x);$

(**c**) $\left(\dfrac{f}{g} \right)'(x) = \dfrac{g(x) \cdot f'(x) - f(x) \cdot g'(x)}{g^2(x)}.$

Proof: Assertion (**a**) is easy and we leave it as an exercise for you.

For **(b)**, we write

$$\lim_{t \to x} \frac{(f \cdot g)(t) - (f \cdot g)(x)}{t - x} = \lim_{t \to x} \left(\frac{(f(t) - f(x)) \cdot g(t)}{t - x} \right.$$
$$\left. + \frac{(g(t) - g(x)) \cdot f(x)}{t - x} \right)$$
$$= \lim_{t \to x} \left(\frac{(f(t) - f(x)) \cdot g(t)}{t - x} \right)$$
$$+ \lim_{t \to x} \left(\frac{(g(t) - g(x)) \cdot f(x)}{t - x} \right)$$
$$= \lim_{t \to x} \left(\frac{(f(t) - f(x))}{t - x} \right) \cdot \left(\lim_{t \to x} g(t) \right)$$
$$+ \lim_{t \to x} \left(\frac{(g(t) - g(x))}{t - x} \right) \cdot \left(\lim_{t \to x} f(x) \right),$$

where we have used Theorem 5.7 about limits. Now the first limit is the derivative of f at x, while the third limit is the derivative of g at x. Also notice that the limit of $g(t)$ equals $g(x)$ by the lemma. The result is that the last line equals

$$f'(x) \cdot g(x) + g'(x) \cdot f(x),$$

as desired.

To prove **(c)**, write

$$\lim_{t \to x} \frac{(f/g)(t) - (f/g)(x)}{t - x} = \lim_{t \to x} \frac{1}{g(t) \cdot g(x)} \left(\frac{f(t) - f(x)}{t - x} \cdot g(x) \right.$$
$$\left. - \frac{g(t) - g(x)}{t - x} \cdot f(x) \right).$$

The proof is now completed by using Theorem 5.7 about limits to evaluate the individual limits in this expression. □

EXAMPLE 6.6 That $f(x) = x$ is differentiable follows from

$$\lim_{t \to x} \frac{t - x}{t - x} = 1.$$

Any constant function is differentiable (with derivative identically zero) by a similar argument. It follows from the theorem that any polynomial function is differentiable.

On the other hand, the continuous function $f(x) = |x|$ is *not* differentiable at the point $x = 0$. This is so because

$$\lim_{t \to 0^-} \frac{|t| - |0|}{t - x} = \lim_{t \to 0^-} \frac{-t - 0}{t - 0} = -1$$

while
$$\lim_{t \to 0^+} \frac{|t| - |0|}{t - x} = \lim_{t \to 0^+} \frac{t - 0}{t - 0} = 1.$$

So the required limit does not exist.   ∎

Since the subject of differential calculus is concerned with learning uses of the derivative, it concentrates on functions which *are* differentiable. One comes away from the subject with the impression that most functions are differentiable except at a few isolated points—as is the case with the function $f(x) = |x|$. Indeed this was what the mathematicians of the nineteenth century thought. Therefore it came as a shock when Karl Weierstrass produced a continuous function that is not differentiable at *any point*. In fact *most* continuous functions are of this nature: their graphs "wiggle" so much that they cannot have a tangent line at any point. Now we turn to an elegant variant of the example of Weierstrass that is due to B. L. van der Waerden (1903–1996).

Theorem 6.7 *Define a function ψ with domain $\mathbb{R}$ by the rule*

$$\psi(x) = \begin{cases} x - n & \text{if } n \leq x < n+1 \text{ and } n \text{ is even} \\ n + 1 - x & \text{if } n \leq x < n+1 \text{ and } n \text{ is odd} \end{cases}$$

for every integer n. The graph of this function is exhibited in Figure 6.3. Then the function

$$f(x) = \sum_{j=1}^{\infty} \left(\frac{3}{4}\right)^j \psi\left(4^j x\right)$$

is continuous at every real x and differentiable at no real x.

The proof of this theorem is intricate, and we put it in an Appendix to this chapter. What is important for you to understand right now is that this remarkable nowhere differentiable function exists, and can be constructed explicitly.

POINT OF CONFUSION 6.8 The proof of Weierstrass's theorem is long, but the idea is simple: the function f is built by piling oscillations on top of oscillations. When the ℓth oscillation is added, it is made very small in size so that it does not cancel the previous oscillations. But it is made very steep so that it will cause the derivative to become large. The practical meaning of the examples of Weierstrass and van der Waerden is that we should realize that differentiability is a very strong and special property of functions. Most continuous functions are not differentiable at any point. Just as an instance, if φ is *any* differentiable function and f is the function from Theorem 6.7, then $h = \varphi + \epsilon f$ will be nowhere differentiable for any $\epsilon > 0$.

When we are proving theorems about continuous functions, we should not think of them in terms of properties of differentiable functions.

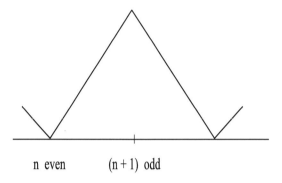

Figure 6.3: The van der Waerden example.

Next we turn to the Chain Rule.

Theorem 6.9 *Let g be a differentiable function on an open interval I and let f be a differentiable function on an open interval that contains the range of g. Then $f \circ g$ is differentiable on the interval I and*

$$(f \circ g)'(x) = f'(g(x)) \cdot g'(x)$$

for each $x \in I$.

Proof: We use the notation Δt to stand for an increment in the variable t. Let us use the symbol $\mathcal{V}(r)$ to stand for any expression which tends to 0 as $\Delta r \to 0$. Fix $x \in I$. Set $r = g(x)$. By hypothesis,

$$\lim_{\Delta r \to 0} \frac{f(r + \Delta r) - f(r)}{\Delta r} = f'(r)$$

or

$$\frac{f(r + \Delta r) - f(r)}{\Delta r} - f'(r) = \mathcal{V}(r)$$

or

$$f(r + \Delta r) = f(r) + \Delta r \cdot f'(r) + \Delta r \cdot \mathcal{V}(r) . \tag{6.9.1}$$

Notice that equation (6.9.1) is valid even when $\Delta r = 0$. Since Δr in equation (6.9.1) can be any small quantity, we set

$$\Delta r = \Delta x \cdot [g'(x) + \mathcal{V}(x)] .$$

Substituting this expression into (6.9.1) and using the fact that $r = g(x)$ yields

$$f(g(x) + \Delta x[g'(x) + \mathcal{V}(x)]) =$$
$$f(r) + (\Delta x \cdot [g'(x) + \mathcal{V}(x)]) \cdot f'(r) + (\Delta x \cdot [g'(x) + \mathcal{V}(x)]) \cdot \mathcal{V}(r)$$
$$= f(g(x)) + \Delta x \cdot f'(g(x)) \cdot g'(x) + \Delta x \cdot \mathcal{V}(x) . \tag{6.9.2}$$

Just as we derived (6.9.1), we may also obtain

$$
\begin{aligned}
g(x + \Delta x) &= g(x) + \Delta x \cdot g'(x) + \Delta x \cdot \mathcal{V}(x) \\
&= g(x) + \Delta x [g'(x) + \mathcal{V}(x)] .
\end{aligned}
$$

We may substitute this equality into the left side of (6.9.2) to obtain

$$
f(g(x + \Delta x)) = f(g(x)) + \Delta x \cdot f'(g(x)) \cdot g'(x) + \Delta x \cdot \mathcal{V}(x) .
$$

With some algebra this can be rewritten as

$$
\frac{f(g(x + \Delta x)) - f(g(x))}{\Delta x} - f'(g(x)) \cdot g'(x) = \mathcal{V}(x) .
$$

But this just says that

$$
\lim_{\Delta x \to 0} \frac{(f \circ g)(x + \Delta x) - (f \circ g)(x)}{\Delta x} = f'(g(x)) \cdot g'(x) .
$$

That is, $(f \circ g)'(x)$ exists and equals $f'(g(x)) \cdot g'(x)$, as desired. □

POINT OF CONFUSION 6.10 It is tempting, in trying to prove the Chain Rule, to reason that

$$
(f \circ g)' = \lim_{\Delta x \to 0} \frac{\Delta f \circ g}{\Delta x} = \lim_{\Delta x \to 0} \frac{\Delta f \circ g}{\Delta g} \cdot \lim_{\Delta x \to 0} \frac{\Delta g}{\Delta x} = f'(g) \cdot g' .
$$

The trouble with this intuitively appealing way of thinking is that Δg could vanish.

A Look Back

1. Say in words what the derivative signifies.

2. Give a verbal explanation of the Chain Rule.

3. What is the significance of the Weierstrass Nowhere Differentiable Function?

4. What properties does a function with positive derivative have?

Exercises

1. For which positive integers k is it true that if $f^k = f \cdot f \cdots f$ is differentiable at x then f is differentiable at x?

2. Let $f(x)$ equal 0 if x is irrational; let $f(x)$ equal $1/q$ if x is a rational number that can be expressed in lowest terms as p/q. Is f differentiable at any x?

3. Formulate notions of "left differentiable" and "right differentiable" for functions defined on suitable half-open intervals. Also formulate definitions of "left continuous" and "right continuous." If you have done things correctly, then you should be able to prove that a left differentiable (right differentiable) function is left continuous (right continuous).

4. Give an example of a function that is infinitely differentiable and that vanishes off the interval $[-a, a]$.

5. Give an example of a function $f : \mathbb{R} \to \mathbb{R}$ which is differentiable at every point but so that the derivative function is not continuous.

6. Prove part (a) of Theorem 6.5.

7. Prove that if f is differentiable on an open interval I and if $f'(x) > 0$ then it is not necessarily the case that f' is positive at points near x.

8. Give an example of a function f on $\mathbb{R}$ so that f' takes on all possible real values.

* 9. Assume that f is a differentiable function on $(-1, 1)$. If the limit $\lim_{x \to 0} f'(x)$ exists then is f' continuous at $x = 0$?

* 10. Prove that the Weierstrass Nowhere Differentiable Function f satisfies

$$\frac{|f(x+h) + f(x-h) - 2f(x)|}{|h|} \leq C|h|$$

for all nonzero h but f is *not* in Lip_1.

* 11. Prove that the Nowhere Differentiable Function constructed in Theorem 6.7 is in Lip_α for all $\alpha < 1$. [Here $f \in \mathrm{Lip}_\alpha$ if $|f(x) - f(t)| \leq C|x - t|^\alpha$ for all x, t.]

* 12. Let $E \subseteq \mathbb{R}$ be a closed set. Fix a nonnegative integer k. Show that there is a function f in $C^k(\mathbb{R})$ (that is, a k-times continuously differentiable function) such that $E = \{x : f(x) = 0\}$.

6.2 The Mean Value Theorem and Applications

Preliminary Remarks

The Mean Value Theorem is a powerful analytic tool. It has intuitive appeal as well. But it is best perceived as a technique of real analysis. In particular, it is Cauchy's version of the Mean Value Theorem that leads to l'Hôpital's Rule.

We begin this section with some remarks about local maxima and minima of functions.

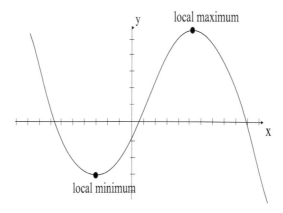

Figure 6.4: Some extrema.

Definition 6.11 Let f be a function with domain (a, b). A point $x \in (a, b)$ is called a *local maximum* for f (we also say that f has a local maximum at x) if there is a $\delta > 0$ such that $f(t) \leq f(x)$ for all $t \in (x - \delta, x + \delta)$. A point $x \in (a, b)$ is called a *local minimum* for f (we also say that f has a local minimum at x) if there is a $\delta > 0$ such that $f(t) \geq f(x)$ for all $t \in (x - \delta, x + \delta)$. See Figure 6.4.

Local minima (plural of minimum) and local maxima (plural of maximum) are referred to collectively as *local extrema*.

Proposition 6.12 (Fermat) *If f is a function with domain (a, b), if f has a local extremum at $x \in (a, b)$, and if f is differentiable at x, then $f'(x) = 0$.*

Proof: Suppose that f has a local minimum at x. Then there is a $\delta > 0$ such that $x - \delta < t < x$ implies $f(t) \geq f(x)$. Hence

$$\frac{f(t) - f(x)}{t - x} \leq 0.$$

Letting $t \to x$, it follows that $f'(x) \leq 0$. Similarly, if $x < t < x + \delta$ for suitable δ, then

$$\frac{f(t) - f(x)}{t - x} \geq 0.$$

It follows that $f'(x) \geq 0$. We must conclude that $f'(x) = 0$.

A similar argument applies if f has a local maximum at x. The proof is therefore complete. □

Before going on to mean value theorems, we provide a striking application of Fermat's proposition:

Theorem 6.13 (Darboux's Theorem) *Let f be a differentiable function on an open interval I. Pick points $s < t$ in I and suppose that $f'(s) < \rho < f'(t)$. Then there is a point u between s and t such that $f'(u) = \rho$.*

Proof: Consider the function $g(x) = f(x) - \rho x$. Then $g'(s) < 0$ and $g'(t) > 0$. Assume for simplicity that $s < t$. The sign of the derivative at s shows that $g(\hat{s}) < g(s)$ for $\hat{s}$ greater than s and near s. The sign of the derivative at t implies that $g(\hat{t}) < g(t)$ for $\hat{t}$ less than t and near t. Thus the minimum of the continuous function g on the compact interval $[s,t]$ must occur at some point u in the interior (s,t). The preceding proposition guarantees that $g'(u) = 0$, or $f'(u) = \rho$ as claimed. □

If f' were a continuous function then the theorem would just be a special instance of the Intermediate Value Property of continuous functions (see Corollary 5.32). But derivatives need not be continuous, as the example

$$f(x) = \begin{cases} x^2 \cdot \sin(1/x) & \text{if} \quad x \neq 0 \\ 0 & \text{if} \quad x = 0 \end{cases}$$

illustrates. Check for yourself that $f'(0)$ exists and vanishes but $\lim_{x \to 0} f'(x)$ does not exist. This example illustrates the significance of the theorem. Since the theorem says that f' will always satisfy the Intermediate Value Property (even when it is not continuous), its discontinuities cannot be of the first kind. In other words:

Proposition 6.14 *If f is a differentiable function on an open interval I then the discontinuities of f' are all of the second kind.*

Next we turn to the simplest form of the Mean Value Theorem.

Theorem 6.15 (Rolle's Theorem) *Let f be a continuous function on the closed interval $[a,b]$ which is differentiable on (a,b). If $f(a) = f(b) = 0$ then there is a point $\xi \in (a,b)$ such that $f'(\xi) = 0$. See Figure 6.5.*

Proof: If f is a constant function, then any point ξ in the interval will do. So assume that f is nonconstant.

Theorem 5.25 guarantees that f will have both a maximum and a minimum in $[a,b]$. If one of these occurs in (a,b), then Proposition 6.12 guarantees that f' will vanish at that point and we are done. If both occur at the endpoints, then all the values of f lie between 0 and 0. In other words f is constant, contradicting our assumption. □

Of course the point ξ in Rolle's theorem need not be unique. If $f(x) = x^3 - x^2 - 2x$ on the interval $[-1,2]$ then $f(-1) = f(2) = 0$ and f' vanishes at *two* points of the interval $(-1,2)$. Refer to Figure 6.6.

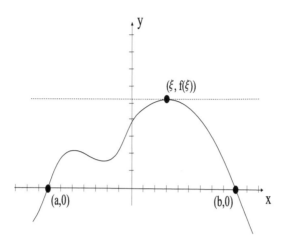

Figure 6.5: Rolle's theorem.

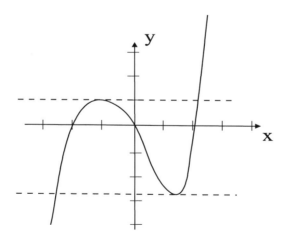

Figure 6.6: An example of Rolle's theorem.

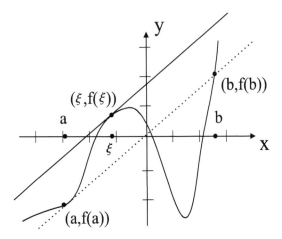

Figure 6.7: The Mean Value Theorem.

If you rotate the graph of a function satisfying the hypotheses of Rolle's theorem, the result suggests that, for any continuous function f on an interval $[a, b]$, differentiable on (a, b), we should be able to relate the slope of the chord connecting $(a, f(a))$ and $(b, f(b))$ with the value of f' at some interior point. That is the content of the standard Mean Value Theorem:

Theorem 6.16 (The Mean Value Theorem) *Let f be a continuous function on the closed interval $[a, b]$ that is differentiable on (a, b). There exists a point $\xi \in (a, b)$ such that*

$$\frac{f(b) - f(a)}{b - a} = f'(\xi).$$

See Figure 6.7.

Proof: Our scheme is to implement the remarks preceding the theorem: we "rotate" the picture to reduce to the case of Rolle's theorem. More precisely, define

$$g(x) = f(x) - \left[f(a) + \frac{f(b) - f(a)}{b - a} \cdot (x - a) \right] \quad \text{if} \quad x \in [a, b].$$

By direct verification, g is continuous on $[a, b]$ and differentiable on (a, b) (after all, g is obtained from f by elementary arithmetic operations). Also $g(a) = g(b) = 0$. Thus we may apply Rolle's theorem to g and we find that there is a $\xi \in (a, b)$ such that $g'(\xi) = 0$. Remembering that x is the variable, we differentiate the formula for g to find that

$$
\begin{aligned}
0 = g'(\xi) &= \left[f'(x) - \frac{f(b) - f(a)}{b - a} \right]\bigg|_{x=\xi} \\
&= \left[f'(\xi) - \frac{f(b) - f(a)}{b - a} \right].
\end{aligned}
$$

As a result,

$$
f'(\xi) = \frac{f(b) - f(a)}{b - a}. \qquad \square
$$

POINT OF CONFUSION 6.17 Suppose that you are driving from point A to point B, that these reference points are 100 miles apart, and that the drive takes you two hours. We may use the Mean Value Theorem to see that, at some point on the trip, we must have been going *exactly* 50 miles per hour.

Corollary 6.18 *If f is a differentiable function on the open interval I and if $f'(x) = 0$ for all $x \in I$ then f is a constant function.*

Proof: If s and t are any two elements of I then the theorem tells us that

$$
f(s) - f(t) = f'(\xi) \cdot (s - t)
$$

for some ξ between s and t. But, by hypothesis, $f'(\xi) = 0$. We conclude that $f(s) = f(t)$. But, since s and t were chosen arbitrarily, we must conclude that f is constant. $\qquad \square$

Corollary 6.19 *If f is differentiable on an open interval I and $f'(x) \geq 0$ for all $x \in I$, then f is *increasing* on I; that is, if $s < t$ are elements of I, then $f(s) \leq f(t)$.*

*If f is differentiable on an open interval I and $f'(x) \leq 0$ for all $x \in I$, then f is *decreasing* on I; that is, if $s < t$ are elements of I, then $f(s) \geq f(t)$.*

Proof: Similar to the preceding corollary. $\qquad \square$

EXAMPLE 6.20 Let us verify that, if f is a differentiable function on $\mathbb{R}$, and if $|f'(x)| \leq 1$ for all x, then $|f(s) - f(t)| \leq |s - t|$ for all real s and t.

In fact, for $s \neq t$ there is a ξ between s and t such that

$$\frac{f(s) - f(t)}{s - t} = f'(\xi).$$

But $|f'(\xi)| \leq 1$ by hypothesis hence

$$\left| \frac{f(s) - f(t)}{s - t} \right| \leq 1$$

or

$$|f(s) - f(t)| \leq |s - t|.$$ ∎

EXAMPLE 6.21 Let us verify that

$$\lim_{x \to +\infty} \left(\sqrt{x + 5} - \sqrt{x} \right) = 0.$$

Here the limit operation means that, for any $\epsilon > 0$, there is an $N > 0$ such that $x > N$ implies that the expression in parentheses has absolute value less than ϵ.

Define $f(x) = \sqrt{x}$ for $x > 0$. Then the expression in parentheses is just $f(x + 5) - f(x)$. By the Mean Value Theorem this equals

$$f'(\xi) \cdot 5$$

for some $x < \xi < x + 5$. But this last expression is

$$\frac{1}{2} \cdot \xi^{-1/2} \cdot 5.$$

By the bounds on ξ, this is

$$\leq \frac{5}{2} x^{-1/2}.$$

Clearly, as $x \to +\infty$, this expression tends to zero.   ∎

A powerful tool in analysis is a generalization of the usual Mean Value Theorem that is due to A. L. Cauchy:

Theorem 6.22 (Cauchy's Mean Value Theorem) *Let f and g be continuous functions on the interval $[a, b]$ which are both differentiable on the interval (a, b). Assume that $g' \neq 0$ on the interval. Then there is a point $\xi \in (a, b)$ such that*

$$\frac{f(b) - f(a)}{g(b) - g(a)} = \frac{f'(\xi)}{g'(\xi)}.$$

Proof: Apply the usual Mean Value Theorem to the function

$$h(x) = g(x) \cdot \{f(b) - f(a)\} - f(x) \cdot \{g(b) - g(a)\} . \qquad \square$$

Clearly the usual Mean Value Theorem (Theorem 6.16) is obtained from Cauchy's by taking $g(x)$ to be the function x.

It is a fact that the standard proof of l'Hôpital's Rule (Guillaume François Antoine de l'Hôpital, Marquis de St.-Mesme, 1661–1704) is obtained by way of Cauchy's Mean Value Theorem. This line of reasoning is explored in the next section.

A Look Back

1. What does Fermat's Test say? Why is it intuitively obvious?

2. What does Rolle's theorem say? Why is its justification clear?

3. What does the Mean Value Theorem say? Why is it a generalization of Rolle's theorem?

4. In what sense is Cauchy's Mean Value Theorem a generalization of the usual Mean Value Theorem?

Exercises

1. Let f be a function that is continuous on $[0, \infty)$ and differentiable on $(0, \infty)$. If $f(0) = 0$ and $|f'(x)| \leq |f(x)|$ for all $x > 0$ then prove that $f(x) = 0$ for all x. [This result is often called *Gronwall's inequality*.]

2. Let f be a continuous function on $[a, b]$ that is differentiable on (a, b). Assume that $f(a) = m$ and that $|f'(x)| \leq K$ for all $x \in (a, b)$. What bound can you then put on the magnitude of $f(b)$?

3. Let f be a differentiable function on an open interval I and assume that f has no local minima nor local maxima on I. Prove that f is either increasing or decreasing on I.

4. Let $0 < \alpha \leq 1$. Prove that there is a constant $C_\alpha > 0$ such that, for $0 < x < 1$, it holds that

$$|\ln x| \leq C_\alpha \cdot x^\alpha$$

for $x \geq 1$. Prove that the constant cannot be taken to be independent of α.

5. Let f be a function that is twice differentiable on $(0, \infty)$ and assume that $f''(x) \geq c > 0$ for all x. Prove that f is not bounded from above.

6. Let f be differentiable on an interval I and $f'(x) > 0$ for all $x \in I$. Does it follow that $(f^2)' > 0$ for all $x \in I$? What additional hypothesis on f will make the conclusion true?

7. Use the Mean Value Theorem to say something about the behavior at ∞ of the function $f(x) = \sqrt{x+1} - \sqrt{x}$.

8. Refer to Exercise **7**. What can you say about the asymptotics at infinity of $\sqrt{x+1}/\sqrt{x}$?

9. Refer to Exercises **7** and **8**. What can you say about the asymptotics at ∞ of $h(x) = \log(x+1) - \log(x)$?

10. Refer to Exercises **7** and **8**. What can you say about the asymptotics at ∞ of $k(x) = e^{x+1} - e^x$?

* 11. Answer Exercise **6** with the exponent 2 replaced by any positive integer exponent.

6.3   More on the Theory of Differentiation

Preliminary Remarks

In this section we study l'Hôpital's Rule and related ideas. There are a number of interesting and nontrivial results, and the exercise is worthwhile.

l'Hôpital's Rule (actually due to his teacher J. Bernoulli (1667-1748)) is a useful device for calculating limits, and a nice application of the Cauchy Mean Value Theorem. Here we present a special case of the theorem.

Theorem 6.23 *Suppose that f and g are differentiable functions on an open interval I and that $p \in I$. If $\lim_{x \to p} f(x) = \lim_{x \to p} g(x) = 0$ and if*

$$\lim_{x \to p} \frac{f'(x)}{g'(x)} \tag{6.23.1}$$

exists and equals a real number ℓ then

$$\lim_{x \to p} \frac{f(x)}{g(x)} = \ell \,.$$

Proof: Fix a real number $a > \ell$. By (6.23.1) there is a number $q > p$ such that, if $p < x < q$, then

$$\frac{f'(x)}{g'(x)} < a \,. \tag{6.23.2}$$

But now, if $p < s < t < q$, then

$$\frac{f(t) - f(s)}{g(t) - g(s)} = \frac{f'(x)}{g'(x)}$$

for some $s < x < t$ (by Cauchy's Mean Value Theorem). It follows then from (6.23.2) that

$$\frac{f(t) - f(s)}{g(t) - g(s)} < a .$$

Now let $s \to p$ and invoke the hypothesis about the zero limit of f and g at p to conclude that

$$\frac{f(t)}{g(t)} \leq a$$

when $p < t < q$. Since a is an arbitrary number to the right of ℓ we conclude that

$$\limsup_{t \to p^+} \frac{f(t)}{g(t)} \leq \ell .$$

Similar arguments show that

$$\liminf_{t \to p^+} \frac{f(t)}{g(t)} \geq \ell ;$$

$$\limsup_{t \to p^-} \frac{f(t)}{g(t)} \leq \ell ;$$

$$\liminf_{t \to p^-} \frac{f(t)}{g(t)} \geq \ell .$$

We conclude that the desired limit exists and equals ℓ. $\qquad\qquad\square$

EXAMPLE 6.24 Let $f(x) = \sin x$ and $g(x) = x$. Then both functions have limit 0 at $p = 0$. So we can apply l'Hôpital's Rule to the limit

$$\lim_{x \to 0} \frac{\sin x}{x} .$$

We see that this limit is equal to

$$\lim_{x \to 0} \frac{[\sin x]'}{x'} = \lim_{x \to 0} \frac{\cos x}{1} = 1 . \qquad\qquad\blacksquare$$

A closely related result, with a similar proof, is this:

Theorem 6.25 *Let I be an open interval and $p \in I$. Suppose that f and g are differentiable functions on $I \setminus \{p\}$. If $\lim_{x \to p} f(x) = \lim_{x \to p} g(x) = \pm\infty$ and if*

$$\lim_{x \to p} \frac{f'(x)}{g'(x)} \qquad\qquad (6.24.1)$$

exists and equals a real number ℓ then

$$\lim_{x \to p} \frac{f(x)}{g(x)} = \ell .$$

EXAMPLE 6.26 Let
$$f(x) = \big|\ln|x|\big|^{(x^2)}.$$
We wish to determine $\lim_{x \to 0} f(x)$. To do so, we define
$$F(x) = \ln f(x) = x^2 \ln\big|\ln|x|\big| = \frac{\ln\big|\ln|x|\big|}{1/x^2}.$$

Notice that both the numerator and the denominator tend to $\pm\infty$ as $x \to 0$. So the hypotheses of l'Hôpital's Rule are satisfied and the limit is

$$\lim_{x \to 0} \frac{\ln\big|\ln|x|\big|}{1/x^2} = \lim_{x \to 0} \frac{1/[x\ln|x|]}{-2/x^3} = \lim_{x \to 0} \frac{-x^2}{2\ln|x|} = 0.$$

Since $\lim_{x \to 0} F(x) = 0$ we may calculate that the original limit has value $\lim_{x \to 0} f(x) = 1$.   ∎

POINT OF CONFUSION 6.27 Be careful when using l'Hôpital's Rule. You cannot apply it to just any quotient. You must verify that the hypotheses are true (i.e., both the numerator and denominator must tend to 0, or else both the numerator and the denominator must tend to $\pm\infty$).

Now we turn our attention to derivatives of inverse functions.

Proposition 6.28 Let f be an invertible function on an interval (a, b) with nonzero derivative at a point $x \in (a, b)$. Let $X = f(x)$. Then $\left(f^{-1}\right)'(X)$ exists and equals $1/f'(x)$.

Proof: Observe that, for $T \neq X$,

$$\frac{f^{-1}(T) - f^{-1}(X)}{T - X} = \frac{1}{\frac{f(t)-f(x)}{t-x}}, \qquad (6.28.1)$$

where $T = f(t)$. Since $f'(x) \neq 0$, the difference quotients for f in the denominator are bounded from zero hence the limit of the formula in (6.28.1) exists. This proves that f^{-1} is differentiable at X and that the derivative at that point equals $1/f'(x)$.   □

EXAMPLE 6.29 We know that the function $f(x) = x^k$, k a positive integer, is one-to-one and differentiable on the interval $(0, 1)$. Moreover the derivative $k \cdot x^{k-1}$ never vanishes on that interval. Therefore the proposition applies and we find for $X \in (0, 1) = f((0, 1))$ that

$$\left(f^{-1}\right)'(X) = \frac{1}{f'(x)} = \frac{1}{f'(X^{1/k})}$$

$$= \frac{1}{k \cdot X^{1-1/k}} = \frac{1}{k} \cdot X^{\frac{1}{k}-1}.$$

In other words,

$$\left(X^{1/k} \right)' = \frac{1}{k} X^{\frac{1}{k}-1}. \qquad \blacksquare$$

We conclude this section by saying a few words about higher derivatives. If f is a differentiable function on an open interval I then we may ask whether the function f' is differentiable. If it is, then we denote its derivative by

$$f'' \quad \text{or} \quad f^{(2)} \quad \text{or} \quad \frac{d^2}{dx^2}f \quad \text{or} \quad \frac{d^2 f}{dx^2},$$

and call it the second derivative of f. Likewise the derivative of the $(k-1)$th derivative, if it exists, is called the kth derivative and is denoted

$$f''\cdots' \quad \text{or} \quad f^{(k)} \quad \text{or} \quad \frac{d^k}{dx^k}f \quad \text{or} \quad \frac{d^k f}{dx^k}.$$

Observe that we cannot even consider whether $f^{(k)}$ exists at a point unless $f^{(k-1)}$ exists in a *neighborhood* of that point.

If f is k times differentiable on an open interval I and if each of the derivatives $f^{(1)}, f^{(2)}, \ldots, f^{(k)}$ is continuous on I then we say that the function f is k *times continuously differentiable* on I. We write $f \in C^k(I)$. Obviously there is some redundancy in this definition since the continuity of $f^{(j-1)}$ follows from the existence of $f^{(j)}$. Thus only the continuity of the last derivative $f^{(k)}$ need be checked. Continuously differentiable functions are useful tools in analysis. We denote the class of k times continuously differentiable functions on I by $C^k(I)$.

For $k = 1, 2, \ldots$ the function

$$f_k(x) = \begin{cases} x^{k+1} & \text{if} \quad x \geq 0 \\ -x^{k+1} & \text{if} \quad x < 0 \end{cases}$$

will be k times continuously differentiable on $\mathbb{R}$ but will fail to be $k+1$ times differentiable at $x = 0$. More dramatically, an analysis similar to the one we used on the Weierstrass Nowhere Differentiable Function shows that the function

$$g_k(x) = \sum_{j=1}^{\infty} \frac{3^j}{4^{j+jk}} \sin(4^j x)$$

is k times continuously differentiable on $\mathbb{R}$ but will not be $k+1$ times differentiable at any point (this function, with $k = 0$, was Weierstrass's original example).

A more refined notion of smoothness/continuity of functions is that of Hölder continuity or Lipschitz continuity (see Section 5.3). If f is a function on an open interval I and if $0 < \alpha \leq 1$ then we say that f satisfies a *Lipschitz condition* of order α on I if there is a constant M such that for all $s, t \in I$ we have

$$|f(s) - f(t)| \leq M \cdot |s - t|^{\alpha}.$$

Such a function is said to be of class $\mathrm{Lip}_\alpha(I)$. Clearly a function of class Lip_α is uniformly continuous on I. For, if $\epsilon > 0$, then we may take $\delta = (\epsilon/M)^{1/\alpha}$: it follows that, for $|s - t| < \alpha$, we have

$$|f(s) - f(t)| \leq M \cdot |s - t|^{\alpha} < M \cdot \epsilon/M = \epsilon.$$

Interestingly, when $\alpha > 1$ the class Lip_α contains only constant functions. For in this instance the inequality

$$|f(s) - f(t)| \leq M \cdot |s - t|^{\alpha}$$

leads to

$$\left| \frac{f(s) - f(t)}{s - t} \right| \leq M \cdot |s - t|^{\alpha - 1}.$$

Because $\alpha - 1 > 0$, letting $s \to t$ yields that $f'(t)$ exists for every $t \in I$ and equals 0. It follows from Corollary 6.18 of the last section that f is constant on I.

Instead of trying to extend the definition of $\mathrm{Lip}_\alpha(I)$ to $\alpha > 1$ it is customary to define classes of functions $C^{k,\alpha}$, for $k = 0, 1, \ldots$ and $0 < \alpha \leq 1$, by the condition that f be of class C^k on I and that $f^{(k)}$ be an element of $\mathrm{Lip}_\alpha(I)$. We leave it as an exercise for you to verify that $C^{k,\alpha} \subseteq C^{\ell,\beta}$ if either $k > \ell$ or both $k = \ell$ and $\alpha \geq \beta$.

A Look Back

1. Give a precise statement of l'Hôpital's Rule.

2. How is the version of l'Hôpital's Rule for numerator and denominator tending to 0 related to the version of the numerator and denominator tending to ∞?

3. Explain pictorially why the formula for the derivative of an inverse function is valid.

4. Explain concretely and in elementary terms why l'Hôpital's Rule should be true for quotients of polynomials.

Exercises

1. Suppose that f is a C^2 function on $\mathbb{R}$ and that if $|f''(x)| \leq C$ for all x. Prove that

$$\left| \frac{f(x + h) + f(x - h) - 2f(x)}{h^2} \right| \leq C.$$

2. Fix a positive integer k. Give an example of two functions f and g neither of which is in C^k but such that $f \cdot g \in C^k$.

3. Fix a positive integer ℓ and define $f(x) = |x|^\ell$. In which class C^k does f lie? In which class $C^{k,\alpha}$ does it lie? [**Hint:** Your answer may depend on the parity of ℓ.]

4. Suppose that f is a continuously differentiable function on an interval I and that $f'(x)$ is never zero. Prove that f is invertible. Then prove that f^{-1} is differentiable. Finally, use the Chain Rule on the identity $f\left(f^{-1}\right) = x$ to derive a formula for $\left(f^{-1}\right)'$.

5. Suppose that a function f on the interval $(0,1)$ has left derivative equal to zero at every point. What conclusion can you draw?

6. We know that the first derivative can be characterized by the Newton quotient. Find an analogous characterization of second derivatives. What about third derivatives?

* **7.** In which class $C^{k,\alpha}$ is the function $x \cdot \ln|x|$ on the interval $[-1/2, 1/2]$? How about the function $x/\ln|x|$?

* **8.** We know that a continuous function on the interval $[0,1]$ can be uniformly approximated by polynomials. But if the function f is continuously differentiable on $[0,1]$, then we can actually say something about the *rate* of approximation. That is, if $\epsilon > 0$ then f can be approximated uniformly within ϵ by a polynomial of degree not greater than $N = N(\epsilon)$. Calculate $N(\epsilon)$.

* **9.** Give an example of a function on $\mathbb{R}$ such that

$$\left| \frac{f(x+h) + f(x-h) - 2f(x)}{h} \right| \leq C$$

for all x and all $h \neq 0$ but f is not in $\mathrm{Lip}_1(\mathbb{R})$.

* **10.** Let f be a differentiable function on an open interval I. Prove that f' is continuous if and only if the inverse image under f' of any point is the intersection of I with a closed set.

* **11.** In the text we give sufficient conditions for the inclusion $C^{k,\alpha} \subseteq C^{\ell,\beta}$. Show that the inclusion is strict if either $k > \ell$ or $k = \ell$ and $\alpha > \beta$.

APPENDIX: Proof of Theorem 6.7 (The Weierstrass Nowhere Differentiable Function)

Since we have not yet discussed series of functions, we take a moment to understand the definition of f. Fix a real x. Then the series becomes a series of numbers, and the jth summand does not exceed $(3/4)^j$ in absolute value. Thus the series converges absolutely; therefore it converges. So it is clear that the displayed formula defines a function of x.

Step I: f is continuous. To see that f is continuous, pick an $\epsilon > 0$. Choose N so large that

$$\sum_{j=N+1}^{\infty} \left(\frac{3}{4}\right)^j < \frac{\epsilon}{4}$$

(we can of course do this because the series $\sum \left(\frac{3}{4}\right)^j$ converges). Now fix x. Observe that, since ψ is continuous and the graph of ψ is composed of segments of slope ± 1, we have

$$|\psi(s) - \psi(t)| \leq |s - t|$$

for all s and t. Moreover $|\psi(s) - \psi(t)| \leq 1$ for all s, t.

For $j = 1, 2, \ldots, N$, pick $\delta_j > 0$ so that, when $|t - x| < \delta_j$, then

$$\left|\psi\left(4^j t\right) - \psi\left(4^j x\right)\right| < \frac{\epsilon}{8}.$$

Let δ be the minimum of $\delta_1, \ldots \delta_N$.

Now, if $|t - x| < \delta$, then

$$
\begin{aligned}
|f(t) - f(x)| &= \left| \sum_{j=1}^{N} \left(\frac{3}{4}\right)^j \cdot \left(\psi(4^j t) - \psi(4^j x)\right) \right. \\
&\qquad \left. + \sum_{j=N+1}^{\infty} \left(\frac{3}{4}\right)^j \cdot \left(\psi(4^j t) - \psi(4^j x)\right) \right| \\
&\leq \sum_{j=1}^{N} \left(\frac{3}{4}\right)^j \left|\left(\psi(4^j t) - \psi(4^j x)\right)\right| \\
&\qquad + \sum_{j=N+1}^{\infty} \left(\frac{3}{4}\right)^j \left|\psi(4^j t) - \psi(4^j x)\right| \\
&\leq \sum_{j=1}^{N} \left(\frac{3}{4}\right)^j \cdot \frac{\epsilon}{8} + \sum_{j=N+1}^{\infty} \left(\frac{3}{4}\right)^j.
\end{aligned}
$$

Here we have used the choice of δ to estimate the summands in the first sum. The first sum is thus less than $\epsilon/2$ (just notice that $\sum_{j=1}^{\infty}(3/4)^j < 4$). The second sum is less than $\epsilon/2$ by the choice of N. Altogether then

$$|f(t) - f(x)| < \epsilon$$

whenever $|t - x| < \delta$. Therefore f is continuous, indeed uniformly so.

Step II: f is nowhere differentiable. Fix x. For $\ell = 1, 2, \ldots$ define $t_\ell = x \pm 4^{-\ell}/2$. We will say whether the sign is plus or minus in a moment (this will depend on the position of x relative to the integers). Then

$$\left| \frac{f(t_\ell) - f(x)}{t_\ell - x} \right| = \left| \frac{1}{t_\ell - x} \left[\sum_{j=1}^{\ell} \left(\frac{3}{4}\right)^j \left(\psi(4^j t_\ell) - \psi(4^j x) \right) \right. \right.$$

$$\left. \left. + \sum_{j=\ell+1}^{\infty} \left(\frac{3}{4}\right)^j \left(\psi(4^j t_\ell) - \psi(4^j x) \right) \right] \right|. \qquad (6.7.1)$$

Notice that, when $j \geq \ell + 1$, then $4^j t_\ell$ and $4^j x$ differ by an even integer. Since ψ has period 2, we find that each of the summands in the second sum is 0. Next we turn to the first sum.

We choose the sign—plus or minus—in the definition of t_ℓ so that there is no integer lying between $4^\ell t_\ell$ and $4^\ell x$. We can do this because the two numbers differ by $1/2$. But then the ℓth summand has magnitude

$$(3/4)^\ell \cdot |4^\ell t_\ell - 4^\ell x| = 3^\ell |t_\ell - x|.$$

On the other hand, the first $\ell - 1$ summands add up to not more than

$$\sum_{j=1}^{\ell-1} \left(\frac{3}{4}\right)^j \cdot |4^j t_\ell - 4^j x| = \sum_{j=1}^{\ell-1} 3^j \cdot 4^{-\ell}/2 \leq \frac{3^\ell - 1}{3 - 1} \cdot 4^{-\ell}/2 \leq 3^\ell \cdot 4^{-\ell-1}.$$

It follows that

$$
\left| \frac{f(t_\ell) - f(x)}{t_\ell - x} \right| \;=\; \frac{1}{|t_\ell - x|} \cdot \left| \sum_{j=1}^{\ell} \left(\frac{3}{4} \right)^j \left(\psi(4^j t_\ell) - \psi(4^j x) \right) \right|
$$

$$
=\; \frac{1}{|t_\ell - x|} \cdot \left| \sum_{j=1}^{\ell-1} \left(\frac{3}{4} \right)^j \left(\psi(4^j t_\ell) - \psi(4^j x) \right) \right.
$$

$$
\left. + \left(\frac{3}{4} \right)^\ell \left(\psi(4^\ell t_\ell) - \psi(4^\ell x) \right) \right|
$$

$$
\geq\; \frac{1}{|t_\ell - x|} \cdot \left| \left(\frac{3}{4} \right)^\ell \psi(4^\ell t_\ell) - \left(\frac{3}{4} \right)^\ell \psi(4^\ell x) \right|
$$

$$
-\; \frac{1}{|t_\ell - x|} \left| \sum_{j=1}^{\ell-1} \left(\frac{3}{4} \right)^j \left(\psi(4^j t_\ell) - \psi(4^j x) \right) \right|
$$

$$
\geq\; 3^\ell - \frac{1}{(4^{-\ell}/2)} \cdot 3^\ell \cdot 4^{-\ell-1}
$$

$$
\geq\; 3^{\ell-1}.
$$

Thus $t_\ell \to x$ but the Newton quotients blow up. Therefore the limit

$$
\lim_{t \to x} \frac{f(t) - f(x)}{t - x}
$$

cannot exist. The function f is not differentiable at x. $\qquad\square$

Chapter 7

The Integral

7.1 Partitions and the Concept of Integral

Preliminary Remarks

It was Bernhard Riemann who came up with the concept of integral that we use today. His concept is based on the idea of a "partition." That is to say, we address the Riemann integral by breaking up the domain of the function. Interestingly, the more advanced idea of the Lebesgue integral is obtained by breaking up the range of the function.

We learn in calculus that it is often useful to think of an integral as representing area. However, this is but one of many important applications of integration theory. The integral is a generalization of the summation process. That is the point of view that we shall take in the present chapter.

Definition 7.1 Let $[a, b]$ be a closed interval in $\mathbb{R}$. A finite, ordered set of points $\mathcal{P} = \{x_0, x_1, x_2, \ldots, x_{k-1}, x_k\}$ such that

$$a = x_0 \leq x_1 \leq x_2 \leq \cdots \leq x_{k-1} \leq x_k = b$$

is called a *partition* of $[a, b]$. Refer to Figure 7.1.

If $\mathcal{P}$ is a partition of $[a, b]$, then we let I_j denote the interval $[x_{j-1}, x_j]$, $j = 1, 2, \ldots, k$. The symbol Δ_j denotes the *length* of I_j. The *mesh* of $\mathcal{P}$, denoted by $m(\mathcal{P})$, is defined to be $\max \Delta_j$.

The points of a partition need not be equally spaced, nor must they be distinct from each other.

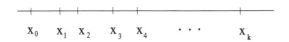

Figure 7.1: A partition.

Figure 7.2: The partition in Example 7.2.

EXAMPLE 7.2 The set $\mathcal{P} = \{0, 1, 1, 9/8, 2, 5, 21/4, 23/4, 6\}$ is a partition of the interval $[0, 6]$ with mesh 3 (because $I_5 = [2, 5]$, with length 3, is the longest interval in the partition). See Figure 7.2. ∎

Definition 7.3 Let $[a, b]$ be an interval and let f be a function with domain $[a, b]$. If $\mathcal{P} = \{x_0, x_1, x_2, \ldots, x_{k-1}, x_k\}$ is a partition of $[a, b]$ and if, for each j, s_j is an element of I_j, then the corresponding *Riemann sum* is defined to be

$$\mathcal{R}(f, \mathcal{P}) = \sum_{j=1}^{k} f(s_j) \Delta_j.$$

EXAMPLE 7.4 Let $f(x) = x^2 - x$ and $[a, b] = [1, 4]$. Define the partition $\mathcal{P} = \{1, 3/2, 2, 7/3, 4\}$ of this interval. Then a Riemann sum for this f and $\mathcal{P}$ is

$$\mathcal{R}(f, \mathcal{P}) = (1^2 - 1) \cdot \frac{1}{2} + ((7/4)^2 - (7/4)) \cdot \frac{1}{2}$$
$$+ ((7/3)^2 - (7/3)) \cdot \frac{1}{3} + (3^2 - 3) \cdot \frac{5}{3}$$
$$= \frac{10103}{864}.$$
∎

POINT OF CONFUSION 7.5 Notice that we have complete latitude in choosing each point s_j from the corresponding interval I_j. While at first confusing, we will find this freedom to be a powerful tool when proving results about the integral.

The first main step in the theory of the Riemann integral is to determine a method for "calculating the limit of the Riemann sums" of a function as the mesh of the partitions tends to zero. There are in fact several means of doing so. We have chosen the simplest one.

Definition 7.6 Let $[a, b]$ be an interval and f a function with domain $[a, b]$. We say that *the Riemann sums of f tend to a limit ℓ as $m(\mathcal{P})$ tends to* 0 if, for any $\epsilon > 0$, there is a $\delta > 0$ such that, if $\mathcal{P}$ is any partition of $[a, b]$ with $m(\mathcal{P}) < \delta$, then $|\mathcal{R}(f, \mathcal{P}) - \ell| < \epsilon$ for every choice of $s_j \in I_j$.

It will turn out to be critical for the success of this definition that we require that *every* partition of mesh smaller than δ satisfy the conclusion of the definition. The theory does not work effectively if for every $\epsilon > 0$ there is a $\delta > 0$ and *some* partition $\mathcal{P}$ of mesh less than δ which satisfies the conclusion of the definition.

Definition 7.7 A function f on a closed interval $[a, b]$ is said to be *Riemann integrable* on $[a, b]$ if the Riemann sums of $\mathcal{R}(f, \mathcal{P})$ tend to a finite limit ℓ as $m(\mathcal{P})$ tends to zero.

The value ℓ of the limit, when it exists, is called the *Riemann integral* of f over $[a, b]$ and is denoted by

$$\int_a^b f(x)\, dx\,.$$

POINT OF CONFUSION 7.8 The value of the integral is well approximated by its Riemann sums. This observation is a useful tool in calculation. We can use Riemann sums to obtain accurate approximations to π and to other important quantities in mathematics. Riemann sums are also powerful devices for studying differential equations and complex analysis.

Remark 7.9 We mention now a useful fact that will be formalized in later sections. Suppose that f is Riemann integrable on $[a, b]$ with the value of the integral being ℓ. Let $\epsilon > 0$. Then, as stated in the definition (with $\epsilon/2$ replacing ϵ), there is a $\delta > 0$ such that, if $\mathcal{Q}$ is a partition of $[a, b]$ of mesh smaller than δ, then $|\mathcal{R}(f, \mathcal{Q}) - \ell| < \epsilon/2$. It follows that, if $\mathcal{P}$ and $\mathcal{P}'$ are partitions of $[a, b]$ of mesh smaller than δ, then

$$|\mathcal{R}(f, \mathcal{P}) - \mathcal{R}(f, \mathcal{P}')| \leq |\mathcal{R}(f, \mathcal{P}) - \ell| + |\ell - \mathcal{R}(f, \mathcal{P}')| < \frac{\epsilon}{2} + \frac{\epsilon}{2} = \epsilon\,.$$

Note, however, that we may choose $\mathcal{P}'$ to equal the partition $\mathcal{P}$. Also we may for each j choose the points s_j, where f is evaluated for the Riemann sum over $\mathcal{P}$, to be a point where f very nearly assumes its supremum on I_j. Likewise we may for each j choose the points s_j', where f is evaluated for the Riemann sum over $\mathcal{P}'$, to be a point where f very nearly assumes its infimum on I_j. It easily follows that, when the mesh of $\mathcal{P}$ is less than δ, then

$$\sum_j \left(\sup_{I_j} f - \inf_{I_j} f \right) \Delta_j \leq \epsilon\,. \tag{7.9.1}$$

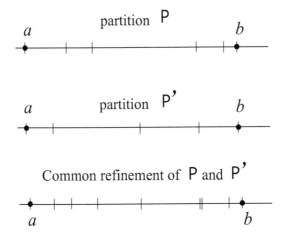

Figure 7.3: The common refinement.

This consequence of integrability will prove useful to us in some of the discussions in this and the next section. In the exercises we shall consider in detail the assertion that integrability implies (7.9.1) and the converse as well.

Definition 7.10 If $\mathcal{P}, \mathcal{P}'$ are partitions of $[a, b]$ then their *common refinement* is the union of all the points of $\mathcal{P}$ and $\mathcal{P}'$. See Figure 7.3.

We record now a technical lemma that will be used in several of the proofs that follow:

Lemma 7.11 *Let f be a function with domain the closed interval $[a, b]$. The Riemann integral*

$$\int_a^b f(x)\, dx$$

exists if and only if, for every $\epsilon > 0$, there is a $\delta > 0$ such that, if $\mathcal{P}$ and $\mathcal{P}'$ are partitions of $[a, b]$ with $m(\mathcal{P}) < \delta$ and $m(\mathcal{P}') < \delta$, then their common refinement $\mathcal{Q}$ has the property that

$$|\mathcal{R}(f, \mathcal{P}) - \mathcal{R}(f, \mathcal{Q})| < \epsilon$$

and   (7.11.1)

$$|\mathcal{R}(f, \mathcal{P}') - \mathcal{R}(f, \mathcal{Q})| < \epsilon.$$

Proof: If f is Riemann integrable, then the assertion of the lemma follows immediately from the definition of the integral.

For the converse note that (7.11.1) certainly implies that, if $\epsilon > 0$, then there is a $\delta > 0$ such that, if $\mathcal{P}$ and $\mathcal{P}'$ are partitions of $[a, b]$ with $m(\mathcal{P}) < \delta$ and $m(\mathcal{P}') < \delta$, then

$$|\mathcal{R}(f, \mathcal{P}) - \mathcal{R}(f, \mathcal{P}')| < \epsilon \qquad (7.11.2)$$

(just use the triangle inequality).

Now, for each $\epsilon_j = 2^{-j}, j = 1, 2, \ldots$, we can choose a $\delta_j > 0$ as in (7.11.2). Let S_j be the *closure* of the set

$$\{\mathcal{R}(f, \mathcal{P}) : m(\mathcal{P}) < \delta_j\}.$$

By the choice of δ_j (and by (7.11.2)), the set S_j is contained in a closed interval of length not greater than $2\epsilon_j$.

On the one hand,

$$\bigcap_j S_j$$

must be nonempty since it is the decreasing intersection of compact sets. On the other hand, the length estimate implies that the intersection must be contained in a closed interval of length 0—that is, the intersection is a point. That point is then the limit of the Riemann sums, that is, it is the value of the Riemann integral. □

The most important, and perhaps the simplest, fact about the Riemann integral is that a large class of familiar functions is Riemann integrable.

Theorem 7.12 *Let f be a continuous function on a nontrivial closed, bounded interval $I = [a, b]$. Then f is Riemann integrable on $[a, b]$.*

Proof: We use the lemma. Given $\epsilon > 0$, choose (by the uniform continuity of f on I—Theorem 5.30) a $\delta > 0$ such that, whenever $|s - t| < \delta$ then

$$|f(s) - f(t)| < \frac{\epsilon}{b - a}. \qquad (7.12.1)$$

Let $\mathcal{P}$ and $\mathcal{P}'$ be any two partitions of $[a, b]$ of mesh smaller than δ. Let $\mathcal{Q}$ be the common refinement of $\mathcal{P}$ and $\mathcal{P}'$.

Now we let I_j denote the intervals arising in the partition $\mathcal{P}$ (and having length Δ_j) and $\widetilde{I}_\ell$ the intervals arising in the partition $\mathcal{Q}$ (and having length $\widetilde{\Delta}_\ell$). Since the partition $\mathcal{Q}$ contains every point of $\mathcal{P}$, plus some additional points as well, every $\widetilde{I}_\ell$ is contained in some I_j. Fix j and consider the expression

$$\left| f(s_j)\Delta_j - \sum_{\widetilde{I}_\ell \subseteq I_j} f(t_\ell)\widetilde{\Delta}_\ell \right|. \qquad (7.12.2)$$

We write

$$\Delta_j = \sum_{\tilde{I}_\ell \subseteq I_j} \tilde{\Delta}_\ell .$$

This equality enables us to rearrange (7.12.2) as

$$\left| f(s_j) \cdot \sum_{\tilde{I}_\ell \subseteq I_j} \tilde{\Delta}_\ell - \sum_{\tilde{I}_\ell \subseteq I_j} f(t_\ell) \tilde{\Delta}_\ell \right|$$

$$= \left| \sum_{\tilde{I}_\ell \subseteq I_j} [f(s_j) - f(t_\ell)] \tilde{\Delta}_\ell \right|$$

$$\leq \sum_{\tilde{I}_\ell \subseteq I_j} |f(s_j) - f(t_\ell)| \tilde{\Delta}_\ell .$$

But each of the points t_ℓ is in the interval I_j, as is s_j. So they differ by less than δ. Therefore, by (7.12.1), the last expression is less than

$$\sum_{\tilde{I}_\ell \subseteq I_j} \frac{\epsilon}{b-a} \tilde{\Delta}_\ell = \frac{\epsilon}{b-a} \sum_{\tilde{I}_\ell \subseteq I_j} \tilde{\Delta}_\ell$$

$$= \frac{\epsilon}{b-a} \cdot \Delta_j .$$

Now we conclude the argument by writing

$$|\mathcal{R}(f, \mathcal{P}) - \mathcal{R}(f, \mathcal{Q})| = \left| \sum_j f(s_j) \Delta_j - \sum_\ell f(t_\ell) \tilde{\Delta}_\ell \right|$$

$$\leq \sum_j \left| f(s_j) \Delta_j - \sum_{\tilde{I}_\ell \subseteq I_j} f(t_\ell) \tilde{\Delta}_\ell \right|$$

$$< \sum_j \frac{\epsilon}{b-a} \cdot \Delta_j$$

$$= \frac{\epsilon}{b-a} \cdot \sum_j \Delta_j$$

$$= \frac{\epsilon}{b-a} \cdot (b-a)$$

$$= \epsilon .$$

The estimate for $|\mathcal{R}(f, \mathcal{P}') - \mathcal{R}(f, \mathcal{Q})|$ is identical and we omit it. The result now follows from Lemma 7.11. $\square$

In the exercises we will ask you to extend the theorem to the case of functions f on $[a, b]$ that are bounded and have finitely many, or even countably many, discontinuities.

We conclude this section by noting an important fact about Riemann integrable functions. A Riemann integrable function on an interval $[a, b]$ *must be bounded*. If it were not, then one could choose the points s_j in the construction of $\mathcal{R}(f, \mathcal{P})$ so that $f(s_j)$ is arbitrarily large, and the Riemann sums would become arbitrarily large, hence cannot converge. You will be asked in the exercises to work out the details of this assertion.

A Look Back

1. What is a partition?

2. What is the mesh of a partition?

3. What is the common refinement of two partitions?

4. What does it mean for the integral of a function f to exist?

5. What is a fairly large class of functions that are Riemann integrable?

Exercises

1. If f is a Riemann integrable function on $[a, b]$, then show that f must be a bounded function.

2. Define the *Dirichlet function* to be

$$f(x) = \begin{cases} 1 & \text{if} \quad x \text{ is rational} \\ 0 & \text{if} \quad x \text{ is irrational} \end{cases}$$

   Prove that the Dirichlet function is not Riemann integrable on the interval $[a, b]$.

3. Define

$$g(x) = \begin{cases} x \cdot \sin(1/x) & \text{if} \quad x \neq 0 \\ 0 & \text{if} \quad x = 0 \end{cases}$$

   Is g Riemann integrable on the interval $[-1, 1]$?

4. Provide the details of the assertion that, if f is Riemann integrable on the interval $[a, b]$ then, for any $\epsilon > 0$, there is a $\delta > 0$ such that, if $\mathcal{P}$ is a partition of mesh less than δ, then

$$\sum_j \left(\sup_{I_j} f - \inf_{I_j} f \right) \Delta_j < \epsilon.$$

   [**Hint:** Follow the scheme presented in Remark 7.9. Given $\epsilon > 0$, choose $\delta > 0$ as in the definition of the integral. Fix a partition $\mathcal{P}$ with mesh smaller than δ. Let $K + 1$ be the number of points in $\mathcal{P}$. Choose points $t_j \in I_j$ so that $|f(t_j) - \sup_{I_j} f| < \epsilon/(2(K + 1))$; also choose points $t'_j \in I_j$ so that $|f(t'_j) - \inf_{I_j} f| < \epsilon/(2(K + 1))$. By applying the definition of the integral to this choice of t_j and t'_j we find that

$$\sum_j \left(\sup_{I_j} f - \inf_{I_j} f \right) \Delta_j < 2\epsilon \,.$$

The result follows.]

5. To what extent is the following statement true? If f is Riemann integrable and bounded from 0 on $[a, b]$ then $1/f$ is Riemann integrable on $[a, b]$.

6. Prove the converse of the statement in Exercise 4. [**Hint:** Note that any Riemann sum over a sufficiently fine partition $\mathcal{P}$ is trapped between the sum in which the infimum is always chosen and the sum in which the supremum is always chosen.]

7. Give an example of a function f such that f^2 is Riemann integrable but f is not.

8. If f is Riemann integrable on the interval $[a, b]$ and if $\mu : [\alpha, \beta] \to [a, b]$ is continuously differentiable then prove that $f \circ \mu$ is Riemann integrable on $[\alpha, \beta]$.

9. Prove that, if f is continuous on the interval $[a, b]$ except at finitely many points and is bounded, then f is Riemann integrable on $[a, b]$.

* 10. Do Exercise **9** with the phrase "finitely many" replaced by "countably many."

* 11. Prove that the Dirichlet function (see Exercise **2**) is the pointwise limit of Riemann integrable functions.

* 12. Show that any Riemann integrable function is the pointwise limit of continuous functions.

* 13. Give an example to show that the composition of Riemann integrable functions need not be Riemann integrable.

7.2   Properties of the Riemann Integral

Preliminary Remarks

Of course the integral is a linear operator on functions, and enjoys thereby a number of useful properties. These include ways in which the integral respects arithmetic operations. We explore these in the present section.

We begin this section with a few elementary properties of the integral that reflect its linear nature.

Theorem 7.13 Let $[a, b]$ be a nonempty interval, let f and g be Riemann integrable functions on the interval, and let α be a real number. Then $f \pm g$ and $\alpha \cdot f$ are integrable and we have

(a) $\int_a^b f(x) \pm g(x)\, dx = \int_a^b f(x)\, dx \pm \int_a^b g(x)\, dx$;

(b) $\int_a^b \alpha \cdot f(x)\, dx = \alpha \cdot \int_a^b f(x)\, dx$.

Proof: For **(a)**, let

$$A = \int_a^b f(x)\, dx$$

and

$$B = \int_a^b g(x)\, dx \,.$$

Let $\epsilon > 0$. Choose a $\delta_1 > 0$ such that if $\mathcal{P}$ is a partition of $[a, b]$ with mesh less than δ_1 then

$$|\mathcal{R}(f, \mathcal{P}) - A| < \frac{\epsilon}{2}\,.$$

Similarly choose a $\delta_2 > 0$ such that if $\mathcal{P}'$ is a partition of $[a, b]$ with mesh less than δ_2 then

$$|\mathcal{R}(f, \mathcal{P}') - B| < \frac{\epsilon}{2}\,.$$

Let $\delta = \min\{\delta_1, \delta_2\}$. If $\mathcal{P}''$ is any partition of $[a, b]$ with $m(\mathcal{P}'') < \delta$ then

$$
\begin{aligned}
|\mathcal{R}(f \pm g, \mathcal{P}'') - (A \pm B)| &= |\mathcal{R}(f, \mathcal{P}'') \pm \mathcal{R}(g, \mathcal{P}'') - (A \pm B)| \\
&\leq |\mathcal{R}(f, \mathcal{P}'') - A| + |\mathcal{R}(g, \mathcal{P}'') - B| \\
&< \frac{\epsilon}{2} + \frac{\epsilon}{2} \\
&= \epsilon.
\end{aligned}
$$

This means that the integral of $f \pm g$ exists and equals $A \pm B$, as we were required to prove.

The proof of **(b)** follows similar lines but is much easier and we leave it as an exercise for you.   □

Theorem 7.14 If c is a point of the interval $[a, b]$ and if f is Riemann integrable on both $[a, c]$ and $[c, b]$ then f is integrable on $[a, b]$ and $\int_a^c f(x)\, dx + \int_c^b f(x)\, dx = \int_a^b f(x)\, dx$.

Proof: Let us write

$$A = \int_a^c f(x)\, dx$$

and

$$B = \int_c^b f(x)\, dx\,.$$

Now pick $\epsilon > 0$. There is a $\delta_1 > 0$ such that if $\mathcal{P}$ is a partition of $[a, c]$ with mesh less than δ_1 then

$$|\mathcal{R}(f, \mathcal{P}) - A| < \frac{\epsilon}{3}\,.$$

Similarly, choose $\delta_2 > 0$ such that if $\mathcal{P}'$ is a partition of $[c, b]$ with mesh less than δ_2 then

$$|\mathcal{R}(f, \mathcal{P}') - B| < \frac{\epsilon}{3}\,.$$

Let M be an upper bound for $|f|$ (recall, from the remark at the end of Section 7.1, that a Riemann integrable function must be bounded). Set $\delta = \min\{\delta_1, \delta_2, \epsilon/(6M)\}$. Now let $\mathcal{V} = \{v_1, \ldots, v_k\}$ be any partition of $[a, b]$ with mesh less than δ. There is a last point v_n which is in $[a, c]$ and a first point v_{n+1} in $[c, b]$. Observe that $\mathcal{P} = \{v_0, \ldots, v_n, c\}$ is a partition of $[a, c]$ with mesh smaller than δ_1 and $\mathcal{P}' = \{c, v_{n+1}, \ldots, v_k\}$ is a partition of $[c, b]$ with mesh smaller than δ_2. Let us rename the elements of $\mathcal{P}$ as $\{p_0, \ldots, p_{n+1}\}$ and the elements of $\mathcal{P}'$ as $\{p'_0, \cdots p'_{k-n+1}\}$. Notice that $p_{n+1} = p'_0 = c$. For each j let s_j be a point chosen in the interval $I_j = [v_{j-1}, v_j]$ from the partition $\mathcal{V}$.

Then we have

$$\left| \mathcal{R}(f, \mathcal{V}) - \left[A + B \right] \right|$$

$$= \left| \left(\sum_{j=1}^n f(s_j)\Delta_j - A \right) + f(s_{n+1})\Delta_{n+1} + \left(\sum_{j=n+2}^k f(s_j)\Delta_j - B \right) \right|$$

$$= \left| \left(\sum_{j=1}^n f(s_j)\Delta_j + f(c) \cdot (c - v_n) - A \right) \right.$$

$$+ \left(f(c) \cdot (v_{n+1} - c) + \sum_{j=n+2}^k f(s_j)\Delta_j - B \right)$$

$$+ \left. \Big(f(s_{n+1}) - f(c) \Big) \cdot (c - v_n) + \Big(f(s_{n+1}) - f(c) \Big) \cdot (v_{n+1} - c) \right|$$

$$\leq \left| \left(\sum_{j=1}^{n} f(s_j)\Delta_j + f(c) \cdot (c - v_n) - A \right) \right|$$

$$+ \left| \left(f(c) \cdot (v_{n+1} - c) + \sum_{j=n+2}^{k} f(s_j)\Delta_j - B \right) \right|$$

$$+ \left| (f(s_{n+1}) - f(c)) \cdot (v_{n+1} - v_n) \right|$$

$$= \left| \mathcal{R}(f, \mathcal{P}) - A \right| + \left| \mathcal{R}(f, \mathcal{P}') - B \right|$$

$$+ \left| (f(s_{n+1}) - f(c)) \cdot (v_{n+1} - v_n) \right|$$

$$< \frac{\epsilon}{3} + \frac{\epsilon}{3} + 2M \cdot \delta$$

$$\leq \epsilon$$

by the choice of δ.

This shows that f is integrable on the entire interval $[a, b]$ and the value of the integral is

$$A + B = \int_a^c f(x)\,dx + \int_c^b f(x)\,dx. \qquad \square$$

Remark 7.15 The last proof illustrates why it is useful to be able to choose the $s_j \in I_j$ arbitrarily. The nub of the proof is to be able to express the integral of f on $[a, b]$, and thus a Riemann sum for that integral, in terms of integrals (and hence Riemann sums) on the two subintervals.

POINT OF CONFUSION 7.16 If we adopt the convention that

$$\int_b^a f(x)\,dx = - \int_a^b f(x)\,dx$$

(which is consistent with the way that the integral was defined in the first place), then Theorem 7.14 is true even when c is not an element of $[a, b]$. For instance, suppose that $c < a < b$. Then, by Theorem 7.14,

$$\int_c^a f(x)\,dx + \int_a^b f(x)\,dx = \int_c^b f(x)\,dx.$$

But this may be rearranged to read

$$\int_a^b f(x)\,dx = - \int_c^a f(x)\,dx + \int_c^b f(x)\,dx = \int_a^c f(x)\,dx + \int_c^b f(x)\,dx.$$

One of the basic tools of analysis is to perform estimates. Thus we require certain fundamental inequalities about integrals. These are recorded in the next theorem.

Theorem 7.17 *Let f and g be integrable functions on a nonempty interval $[a, b]$. Then*

(i) $\left| \int_a^b f(x) \, dx \right| \leq \int_a^b |f(x)| \, dx;$

(ii) *If $f(x) \leq g(x)$ for all $x \in [a, b]$ then $\int_a^b f(x) \, dx \leq \int_a^b g(x) \, dx.$*

Proof: If $\mathcal{P}$ is any partition of $[a, b]$ then

$$|\mathcal{R}(f, \mathcal{P})| \leq \mathcal{R}(|f|, \mathcal{P}).$$

Assertion **(i)** follows.
   Next, for part **(ii)**,
$$\mathcal{R}(f, \mathcal{P}) \leq \mathcal{R}(g, \mathcal{P}).$$

This inequality implies the second assertion.   □

Another fundamental operation in the theory of the integral is "change of variable" (sometimes called the "u-substitution" in calculus books). We next turn to a careful formulation and proof of this operation. First we need a lemma:

Lemma 7.18 *If f is Riemann integrable on an interval $[a, b]$ with values in $[c, d]$, and if $\phi : [c, d] \to \mathbb{R}$ is continuously differentiable, then $\phi \circ f$ is Riemann integrable.*

Proof: The proof is complicated, and we omit the details. See [KRA5, Ch. 7] for the full story.   □

Corollary 7.19 *If f and g are Riemann integrable on $[a, b]$, then so is the function $f \cdot g$.*

Proof: By Theorem 7.13, $f + g$ is integrable. By the lemma, $(f + g)^2 = f^2 + 2f \cdot g + g^2$ is integrable. But the lemma also implies that f^2 and g^2 are integrable (here we use the function $\phi(x) = x^2$). It results, by subtraction, that $2 \cdot f \cdot g$ is integrable. Hence $f \cdot g$ is integrable.   □

Theorem 7.20 Let f be an integrable function on an interval $[a, b]$ of positive length. Let ψ be a continuously differentiable function from another interval $[\alpha, \beta]$ of positive length into $[a, b]$. Assume that ψ is increasing, one-to-one, and onto. Then

$$\int_a^b f(x)\, dx = \int_\alpha^\beta f(\psi(x)) \cdot \psi'(x)\, dx\,.$$

Proof: Since f is integrable, its absolute value is bounded by some number M. Fix $\epsilon > 0$. Since ψ' is continuous on the compact interval $[\alpha, \beta]$, it is uniformly continuous (Theorem 5.30). Hence we may choose $\delta > 0$ so small that if $|s - t| < \delta$ then $|\psi'(s) - \psi'(t)| < \epsilon/(M \cdot (\beta - \alpha))$. If $\mathcal{P} = \{p_0, \ldots, p_k\}$ is any partition of $[a, b]$ then there is an associated partition $\widetilde{\mathcal{P}} = \{\psi^{-1}(p_0), \ldots, \psi^{-1}(p_k)\}$ of $[\alpha, \beta]$. For simplicity denote the points of $\widetilde{\mathcal{P}}$ by $\widetilde{p}_j$. Let us choose the partition $\mathcal{P}$ so fine that the mesh of $\widetilde{\mathcal{P}}$ is less than δ. If t_j are points of $I_j = [p_{j-1}, p_j]$ then there are corresponding points $s_j = \psi^{-1}(t_j)$ of $\widetilde{I}_j = [\widetilde{p}_{j-1}, \widetilde{p}_j]$. Then we have

$$
\begin{aligned}
\sum_{j=1}^k f(t_j)\Delta_j &= \sum_{j=1}^k f(t_j)(p_j - p_{j-1}) \\
&= \sum_{j=1}^k f(\psi(s_j))(\psi(\widetilde{p}_j) - \psi(\widetilde{p}_{j-1})) \\
&= \sum_{j=1}^k f(\psi(s_j))\psi'(u_j)(\widetilde{p}_j - \widetilde{p}_{j-1}),
\end{aligned}
$$

where we have used the Mean Value Theorem in the last line to find each u_j. Our problem at this point is that $f \circ \psi$ and ψ' are evaluated at different points. So we must do some estimation to correct that problem.

The last displayed line equals

$$\sum_{j=1}^k f(\psi(s_j))\psi'(s_j)(\widetilde{p}_j - \widetilde{p}_{j-1}) + \sum_{j=1}^k f(\psi(s_j))\left(\psi'(u_j) - \psi'(s_j)\right)(\widetilde{p}_j - \widetilde{p}_{j-1})\,.$$

The first sum is a Riemann sum for $f(\psi(x)) \cdot \psi'(x)$ and the second sum is an error term. Since the points u_j and s_j are elements of the same interval $\widetilde{I}_j$ of length less than δ, we conclude that $|\psi'(u_j) - \psi'(s_j)| < \epsilon/(M \cdot |\beta - \alpha|)$. Thus the error term in absolute value does not exceed

$$\sum_{j=1}^k M \cdot \frac{\epsilon}{M \cdot |\beta - \alpha|} \cdot (\widetilde{p}_j - \widetilde{p}_{j-1}) = \frac{\epsilon}{\beta - \alpha} \sum_{j=0}^k (\widetilde{p}_j - \widetilde{p}_{j-1}) = \epsilon\,.$$

This shows that every Riemann sum for f on $[a, b]$ with sufficiently small mesh corresponds to a Riemann sum for $f(\psi(x)) \cdot \psi'(x)$ on $[\alpha, \beta]$ plus an error term of size less than ϵ. A similar argument shows that every Riemann sum for $f(\psi(x)) \cdot \psi'(x)$ on $[\alpha, \beta]$ with sufficiently small mesh corresponds to a Riemann sum for f on $[a, b]$ plus an error term of magnitude less than ϵ. The conclusion is then that the integral of f on $[a, b]$ (which exists by hypothesis) and the integral of $f(\psi(x)) \cdot \psi'(x)$ on $[\alpha, \beta]$ (which exists by the corollary to the lemma) agree.   □

We conclude this section with the very important Fundamental Theorem of Calculus.

Theorem 7.21 (The Fundamental Theorem of Calculus) *Let f be a continuous function on the interval $[a, b]$. For $x \in [a, b]$ we define*

$$F(x) = \int_a^x f(s)\, ds\,.$$

For any $x \in (a, b)$ we then have

$$F'(x) = f(x)\,.$$

Proof: Fix $x \in (a, b)$. Let $\epsilon > 0$. Choose, by the continuity of f at x, a $\delta > 0$ such that $|s - x| < \delta$ implies $|f(s) - f(x)| < \epsilon$. We may assume that $\delta < \min\{x - a, b - x\}$. If $|t - x| < \delta$ then

$$\left| \frac{F(t) - F(x)}{t - x} - f(x) \right| = \left| \frac{\int_a^t f(s)\, ds - \int_a^x f(s)\, ds}{t - x} - f(x) \right|$$

$$= \left| \frac{\int_x^t f(s)\, ds}{t - x} - \frac{\int_x^t f(x)\, ds}{t - x} \right|$$

$$= \left| \frac{\int_x^t (f(s) - f(x))\, ds}{t - x} \right|\,.$$

Notice that we rewrote $f(x)$ as the integral with respect to a dummy variable s over an interval of length $|t - x|$ divided by $(t - x)$. Assume for the moment that $t > x$. Then the last line is dominated by

$$\frac{\int_x^t |f(s) - f(x)|\, ds}{t - x} \quad \leq \quad \frac{\int_x^t \epsilon\, ds}{t - x}$$

$$= \quad \epsilon.$$

A similar estimate holds when $t < x$ (simply reverse the limits of integration).

This shows that
$$\lim_{t \to x} \frac{F(t) - F(x)}{t - x}$$
exists and equals $f(x)$. Thus $F'(x)$ exists and equals $f(x)$. □

In the exercises we shall consider how to use the theory of one-sided limits to make the conclusion of the Fundamental Theorem true on the entire interval $[a, b]$. We conclude with

Corollary 7.22 *If f is a continuous function on $[a, b]$ and if G is any continuously differentiable function on $[a, b]$ whose derivative equals f on (a, b) then*
$$\int_a^b f(x)\, dx = G(b) - G(a).$$

Proof: Define F as in the theorem. Since F and G have the same derivative on (a, b), they differ by a constant (Corollary 6.18). Then
$$\int_a^b f(x)\, dx = F(b) = F(b) - F(a) = G(b) - G(a)$$
as desired. □

A Look Back

1. How does change of variable in an integral work?
2. How are we allowed to break up the domain of integration of a function?
3. Why is the Fundamental Theorem of Calculus important?
4. Why are there two different statements of the Fundamental Theorem of Calculus?

Exercises

1. Imitate the proof of the Fundamental Theorem of Calculus in this section to show that, if f is continuous on $[a, b]$ and if we define
$$F(x) = \int_a^x f(t)\, dt,$$
then the one-sided derivative $F'(a)$ exists and equals $f(a)$ in the sense that
$$\lim_{t \to a^+} \frac{F(t) - F(a)}{t - a} = f(a).$$

Formulate and prove an analogous statement for the one-sided derivative of F at b.

2. Let f be a bounded function on an unbounded interval of the form $[A, \infty)$. We say that f is integrable on $[A, \infty)$ if f is integrable on every compact subinterval of $[A, \infty)$ and

$$\lim_{B \to +\infty} \int_A^B f(x)\, dx$$

exists and is finite.

Assume that f is Riemann integrable on $[1, N]$ for every $N > 1$ and that f is decreasing. Show that f is Riemann integrable on $[1, \infty)$ if and only if $\sum_{j=1}^{\infty} f(j)$ is finite.

Suppose that g is nonnegative and integrable on $[1, \infty)$. If $0 \le |f(x)| \le g(x)$ for $x \in [1, \infty)$ and f is integrable on compact subintervals of $[1, \infty)$, then prove that f is integrable on $[1, \infty)$.

3. Let f be a function on an interval of the form $(a, b]$ such that f is integrable on compact subintervals of $(a, b]$. If

$$\lim_{\epsilon \to 0^+} \int_{a+\epsilon}^b f(x)\, dx$$

exists and is finite then we say that f is integrable on $(a, b]$. Prove that, if we restrict attention to bounded f, then in fact this definition gives rise to no new integrable functions. However, there are unbounded functions that can now be integrated. Give an example.

Give an example of a function g that is integrable by the definition in the preceding paragraph but is such that $|g|$ is not integrable.

4. Suppose that f is a continuous, nonnegative function on the interval $[0, 1]$. Let M be the maximum of f on the interval. Prove that

$$\lim_{n \to \infty} \left[\int_0^1 f(t)^n\, dt \right]^{1/n} = M.$$

5. Let f be a continuously differentiable function on the interval $[0, 2\pi]$. Further assume that $f(0) = f(2\pi)$ and $f'(0) = f'(2\pi)$. For $n \in \mathbb{N}$ define

$$\widehat{f}(n) = \frac{1}{2\pi} \int_0^{2\pi} f(x) \sin nx\, dx.$$

Prove that

$$\sum_{n=1}^{\infty} |\widehat{f}(n)|^2$$

converges. [**Hint:** Use integration by parts to obtain a favorable estimate on $|\widehat{f}(n)|$.]

6. Prove part (b) of Theorem 7.13.

7. Refer to Exercise **10**. Prove that if f is twice continuously differentiable on $\mathbb{R}$ and $f \equiv 0$ off the interval $[-1, 1]$, then

$$|\widehat{f}(n)| \leq C \cdot n^{-2}.$$

* **8.** Prove that

$$\lim_{\eta \to 0+} \int_{\eta}^{1/\eta} \frac{\cos(2r) - \cos r}{r} \, dr$$

exists and is finite.

* **9.** Suppose that f is a Riemann integrable function on the interval $[0, 1]$. Let $\epsilon > 0$. Show that there is a polynomial p so that

$$\int_0^1 |f(x) - p(x)| \, dx < \epsilon.$$

* **10.** Let $f_1, f_2, \ldots$ be Riemann integrable functions on $[0, 1]$. Suppose that $f_1(x) \geq f_2(x) \geq \cdots$ for every x and that $\lim_{j \to \infty} f_j(x) \equiv f(x)$ exists and is finite for every x. Is it the case that f is Riemann integrable?

* **11.** Refer to Exercise **5** for terminology. Prove that $\lim_{n \to \pm\infty} |\widehat{f}(n)| = 0$.

Chapter 8

Sequences and Series of Functions

8.1 Convergence of a Sequence of Functions

Preliminary Remarks

Sequences and series of functions play a pivotal role in modern mathematics, and also in engineering and physics. The theory of Fourier series, just as an example, was invented in the early nineteenth century as a means of decomposing a fairly arbitrary function into simple units (namely sines and cosines). The more modern theory of wavelets takes the Fourier theory to new heights of beauty and power.

A *sequence of functions* is usually written

$$f_1, f_2, \ldots \quad \text{or} \quad \{f_j\}_{j=1}^{\infty} .$$

We will generally assume that the functions f_j all have the same domain S.

Definition 8.1 A sequence of functions $\{f_j\}_{j=1}^{\infty}$ with domain $S \subseteq \mathbb{R}$ is said to *converge pointwise* to a limit function f on S if, for each $x \in S$, the sequence of numbers $\{f_j(x)\}$ converges to $f(x)$.

EXAMPLE 8.2 Define $f_j(x) = x^j$ with domain $S = \{x : 0 \le x \le 1\}$. If $0 \le x < 1$ then $f_j(x) \to 0$. However, $f_j(1) \to 1$. Therefore the sequence f_j

converges pointwise to the function

$$f(x) = \begin{cases} 0 & \text{if} \quad 0 \le x < 1 \\ 1 & \text{if} \quad x = 1 . \end{cases}$$

See Figure 8.1. We see that, even though the f_j are each continuous, the limit function f is not.   ∎

Here are some of the basic questions that we must ask about a sequence of functions f_j that converges to a function f on a domain S:

(1) If the functions f_j are continuous, then is f continuous?

(2) If the functions f_j are integrable on an interval I, then is f integrable on I?

(3) If f is integrable on I, then does the sequence $\int_I f_j(x)\,dx$ converge to $\int_I f(x)\,dx$?

(4) If the functions f_j are differentiable, then is f differentiable?

(5) If f is differentiable, then does the sequence f'_j converge to f'?

We see from Example 8.1 that the answer to the first question is "no": Each of the f_j is continuous but f certainly is not. It turns out that, in order to obtain a favorable answer to our questions, we must consider a stricter notion of convergence of functions. This motivates the next definition.

Definition 8.3 Let f_j be a sequence of functions on a domain S. We say that the functions f_j *converge uniformly* to f if, given $\epsilon > 0$, there is an $N > 0$ such that, for any $j > N$ and any $x \in S$, it holds that $|f_j(x) - f(x)| < \epsilon$.

POINT OF CONFUSION 8.4 Notice that the special feature of uniform convergence is that the rate at which $f_j(x)$ converges is independent of $x \in S$. In Example 8.1, $f_j(x)$ is converging very rapidly to zero for x near zero but arbitrarily slowly to zero for x near 1—see Figure 8.1. In the next example we shall prove this assertion rigorously.

EXAMPLE 8.5 The sequence $f_j(x) = x^j$ does *not* converge uniformly to the limit function

$$f(x) = \begin{cases} 0 & \text{if} \quad 0 \le x < 1 \\ 1 & \text{if} \quad x = 1 \end{cases}$$

on the domain $S = [0, 1]$. In fact it does not even do so on the smaller domain $[0, 1)$. To see this, notice that, no matter how large j is we have, by the Mean Value Theorem, that

$$f_j(1) - f_j(1 - 1/(2j)) = \frac{1}{2j} \cdot f'_j(\xi)$$

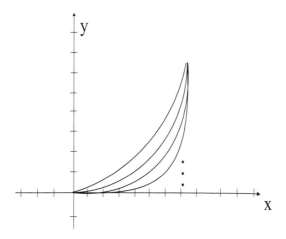

Figure 8.1: The sequence $\{x^j\}$.

for some ξ between $1 - 1/(2j)$ and 1. But $f_j'(x) = j \cdot x^{j-1}$ hence $|f_j'(\xi)| < j$ and we conclude that

$$|f_j(1) - f_j(1 - 1/(2j))| < \frac{1}{2}$$

or

$$f_j(1 - 1/(2j)) > f_j(1) - \frac{1}{2} = \frac{1}{2}.$$

In conclusion, no matter how large j, there will be values of x (namely, $x = 1 - 1/(2j)$) at which $f_j(x)$ is at least distance $1/2$ from the limit 0. We conclude that the convergence is not uniform. ∎

Theorem 8.6 *If f_j are continuous functions on a set S that converge uniformly on S to a function f then f is also continuous.*

Proof: Let $\epsilon > 0$. Fix an integer N so large that, if $j > N$, then $|f_j(x) - f(x)| < \epsilon/3$ for all $x \in S$. Fix $c \in S$. Choose $\delta > 0$ so small that if $x \in S$ and $|x - c| < \delta$ then $|f_N(x) - f_N(c)| < \epsilon/3$. For such x we have

$$
\begin{aligned}
|f(x) - f(c)| &\leq |f(x) - f_N(x)| + |f_N(x) - f_N(c)| + |f_N(c) - f(c)| \\
&< \frac{\epsilon}{3} + \frac{\epsilon}{3} + \frac{\epsilon}{3}
\end{aligned}
$$

by the way that we chose N and δ. But the last line sums to ϵ, proving that f is continuous at c. Since $c \in S$ was chosen arbitrarily, we are done. □

EXAMPLE 8.7 Define functions

$$f_j(x) = \begin{cases} 0 & \text{if} \quad x = 0 \\ j & \text{if} \quad 0 < x \le 1/j \\ 0 & \text{if} \quad 1/j < x \le 1 \end{cases}$$

for $j = 2, 3, \ldots$. Then $\lim_{j \to \infty} f_j(x) = 0 \equiv f(x)$ for all x in the interval $I = [0, 1]$. However

$$\int_0^1 f_j(x)\, dx = \int_0^{1/j} j\, dx = 1$$

for every j. Thus the f_j converge to the integrable limit function $f(x) \equiv 0$ (with integral 0), but their integrals do not converge to the integral of f. ∎

EXAMPLE 8.8 Let $q_1, q_2, \ldots$ be an enumeration of the rationals in the interval $I = [0, 1]$. Define functions

$$f_j(x) = \begin{cases} 1 & \text{if} \quad x \in \{q_1, q_2, \ldots, q_j\} \\ 0 & \text{if} \quad x \notin \{q_1, q_2, \ldots, q_j\} \end{cases}$$

Then the functions f_j converge pointwise to the Dirichlet function f which is equal to 1 on the rationals and 0 on the irrationals. Each of the functions f_j has integral 0 on I. But the function f is not integrable on I. ∎

The last two examples show that something more than pointwise convergence is needed in order for the integral to respect the limit process.

Theorem 8.9 *Let f_j be integrable functions on a nontrivial bounded interval $[a, b]$ and suppose that the functions f_j converge uniformly to the limit function f. Then f is integrable on $[a, b]$ and*

$$\lim_{j \to \infty} \int_a^b f_j(x)\, dx = \int_a^b f(x)\, dx\,.$$

Proof: Pick $\epsilon > 0$. Choose N so large that if $j > N$ then $|f_j(x) - f(x)| < \epsilon/[6(b-a)]$ for all $x \in [a, b]$. Notice that, if $j, k > N$, then

$$\left| \int_a^b f_j(x)\, dx - \int_a^b f_k(x)\, dx \right| \le \int_a^b |f_j(x) - f_k(x)|\, dx\,. \qquad (8.9.1)$$

But $|f_j(x) - f_k(x)| \le |f_j(x) - f(x)| + |f(x) - f_k(x)| < \epsilon/[3(b-a)]$. Therefore line (8.9.1) does not exceed

$$\int_a^b \frac{\epsilon}{3(b-a)}\, dx = \frac{\epsilon}{3}\,.$$

Thus the numbers $\int_a^b f_j(x)\,dx$ form a Cauchy sequence. Let the limit of this sequence be called A. Notice that, if we let $k \to \infty$ in the inequality

$$\left| \int_a^b f_j(x)\,dx - \int_a^b f_k(x)\,dx \right| \le \frac{\epsilon}{3},$$

then we obtain

$$\left| \int_a^b f_j(x)\,dx - A \right| \le \frac{\epsilon}{3}$$

for all $j > N$. This estimate will be used below.

By hypothesis there is a $\delta > 0$ such that, if $\mathcal{P} = \{p_1, \ldots, p_k\}$ is a partition of $[a, b]$ with $m(\mathcal{P}) < \delta$, then

$$\left| \mathcal{R}(f_N, \mathcal{P}) - \int_a^b f_N(x)\,dx \right| < \frac{\epsilon}{3}.$$

But then, for such a partition, we have

$$
\begin{aligned}
|\mathcal{R}(f, \mathcal{P}) - A| \ \le\ & \left| \mathcal{R}(f, \mathcal{P}) - \mathcal{R}(f_N, \mathcal{P}) \right| + \left| \mathcal{R}(f_N, \mathcal{P}) - \int_a^b f_N(x)\,dx \right| \\
& + \left| \int_a^b f_N(x)\,dx - A \right|.
\end{aligned}
$$

We have already noted that, by the choice of N, the third term on the right does not exceed $\epsilon/3$. The second term is smaller than $\epsilon/3$ by the way that we chose the partition $\mathcal{P}$. It remains to examine the first term. Now

$$
\begin{aligned}
\left| \mathcal{R}(f, \mathcal{P}) - \mathcal{R}(f_N, \mathcal{P}) \right| &= \left| \sum_{j=1}^k f(s_j)\Delta_j - \sum_{j=1}^k f_N(s_j)\Delta_j \right| \\
&\le \sum_{j=1}^k \left| f(s_j) - f_N(s_j) \right| \Delta_j \\
&< \sum_{j=1}^k \frac{\epsilon}{6(b-a)} \Delta_j \\
&= \frac{\epsilon}{6(b-a)} \sum_{j=1}^k \Delta_j \\
&= \frac{\epsilon}{6}.
\end{aligned}
$$

Therefore $|\mathcal{R}(f, \mathcal{P}) - A| < \epsilon$ when $m(\mathcal{P}) < \delta$. This shows that the function f is integrable on $[a, b]$ and has integral with value A. $\qquad\square$

POINT OF CONFUSION 8.10 We have seen a few instances now in which the concept of uniform convergence addresses significant issues of convergence of a sequence of functions. This is an important idea for you to master and to be able to use easily.

We have succeeded in answering questions (**1**) and (**2**) that were raised at the beginning of the section. In the next section we will answer questions (**3**), (**4**), and (**5**).

A Look Back

1. What does it mean for a sequence of functions to converge pointwise?

2. What does it mean for a sequence of functions to converge uniformly?

3. What condition on a sequence of integrable functions will guarantee that their limit function will be integrable?

4. Is it true that the limit of a sequence of continuous functions is continuous?

Exercises

1. If $f_j \to f$ uniformly on a domain S and if f_j, f never vanish on S then does it follow that the functions $1/f_j$ converge uniformly to $1/f$ on S?

2. Write out the first five partial sums for the series

$$\sum_{j=1}^{\infty} \frac{\sin^3 j}{j^2}.$$

3. Write a series of polynomials that converges to $f(x) = \sin x^2$. Can you prove that it converges?

4. Write a series of trigonometric functions that converges to $f(x) = x$. Can you prove that it converges?

5. Write a series of piecewise linear functions that converges to $f(x) = x^2$ on the interval $[0, 1]$. Can you prove that it converges?

6. Write a series of functions that converges pointwise on $[0, 1]$ but does not converge uniformly on any proper subinterval. [**Hint:** First consider a sequence.]

7. Show that if $\sum_j f_j'$ converges uniformly on $[0, 1]$ (where the prime stands for the derivative), and if $f_j(0) = 0$ for all j, then $\sum_j f_j$ converges uniformly on compact sets.

8. TRUE or FALSE: If $\sum_j f_j$ converges absolutely and uniformly and $\sum_j g_j$ converges absolutely and uniformly on a compact interval $[a, b]$, then so does $\sum_j f_j g_j$.

9. If a power series $\sum a_j x^j$ converges at each point $x = 1$, $x = 2$, $x = 3$, etc., then show that the series converges uniformly on each interval of the form $[-N, N]$.

* 10. Give an example of a Taylor series that converges uniformly on compact sets to its limit function.

* 11. A Taylor series will never converge only pointwise. Explain.

8.2 More on Uniform Convergence

Preliminary Remarks

As we have seen, uniform convergence is a very powerful idea for guaranteeing that the limit of a sequence of functions is well behaved. In the present section we will see some very explicit and useful applications and extensions of this idea.

In general, limits do not commute. Since the integral is defined with a limit, and since we saw in the last section that integrals do not always respect limits of functions, we know some concrete instances of non-commutation of limits. The fact that continuity is defined with a limit, and that the limit of continuous functions need not be continuous, gives even more examples of situations in which limits do not commute. Let us now turn to a situation in which limits *do* commute:

Theorem 8.11 *Fix a set S and a point $s \in S$. Assume that the functions f_j converge uniformly on the domain $S \setminus \{s\}$ to a limit function f. Suppose that each function $f_j(x)$ has a limit as $x \to s$. Then f itself has a limit as $x \to s$ and*

$$\lim_{x \to s} f(x) = \lim_{j \to \infty} \lim_{x \to s} f_j(x).$$

Because of the way that f is defined, we may rewrite this conclusion as

$$\lim_{x \to s} \lim_{j \to \infty} f_j(x) = \lim_{j \to \infty} \lim_{x \to s} f_j(x).$$

In other words, the limits $\lim_{x \to s}$ and $\lim_{j \to \infty}$ commute.

Proof: Let $\alpha_j = \lim_{x \to s} f_j(x)$. Let $\epsilon > 0$. There is a number $N > 0$ (independent of $x \in S \setminus \{s\}$) such that $j > N$ implies that $|f_j(x) - f(x)| < \epsilon/4$. Fix $j, k > N$. Choose $\delta > 0$ such that $0 < |x - s| < \delta$ implies both that $|f_j(x) - \alpha_j| < \epsilon/4$ and $|f_k(x) - \alpha_k| < \epsilon/4$. Then

$$|\alpha_j - \alpha_k| \le |\alpha_j - f_j(x)| + |f_j(x) - f(x)| + |f(x) - f_k(x)| + |f_k(x) - \alpha_k| .$$

The first and last expressions are less than $\epsilon/4$ by the choice of x. The middle two expressions are less than $\epsilon/4$ by the choice of N. We conclude that the sequence α_j is Cauchy. Let α be the limit of that sequence.

Letting $k \to \infty$ in the inequality

$$|\alpha_j - \alpha_k| < \epsilon$$

that we obtained above yields

$$|\alpha_j - \alpha| \leq \epsilon$$

for $j > N$. Now, with δ as above and $0 < |x - s| < \delta$, we have

$$|f(x) - \alpha| \leq |f(x) - f_j(x)| + |f_j(x) - \alpha_j| + |\alpha_j - \alpha| .$$

By the choices we have made, the first term is less than $\epsilon/4$, the second is less than $\epsilon/2$, and the third is less than or equal to ϵ. Altogether, if $0 < |x - s| < \delta$ then $|f(x) - \alpha| < 2\epsilon$. This is the desired conclusion.   □

POINT OF CONFUSION 8.12 Once again we see that uniform convergence is the key to understanding a tricky situation involving limits. It will be a recurring idea in the rest of this text.

Parallel with our notion of Cauchy sequence of numbers, we have a concept of Cauchy sequence of functions in the uniform sense:

Definition 8.13 A sequence of functions f_j on a domain S is called *a uniformly Cauchy sequence* if, for each $\epsilon > 0$, there is an $N > 0$ such that, if $j, k > N$, then
$$|f_j(x) - f_k(x)| < \epsilon \quad \forall x \in S .$$

POINT OF CONFUSION 8.14 We see that a uniformly Cauchy sequence is one for which the functions get closer together at a uniform rate across the common domain of the functions. This is consistent with our notion of uniform convergence discussed earlier.

Proposition 8.15 A sequence of functions f_j is uniformly Cauchy on a domain S if and only if the sequence converges uniformly to a limit function f on the domain S.

Proof: The proof is straightforward and is assigned as an exercise.   □

We will use the last two results in our study of the limits of differentiable functions. First we consider an example.

EXAMPLE 8.16 Define the function

$$f_j(x) = \begin{cases} 0 & \text{if} \quad x \le 0 \\ jx^2 & \text{if} \quad 0 < x \le 1/(2j) \\ x - 1/(4j) & \text{if} \quad 1/(2j) < x < \infty \end{cases}$$

We leave it as an exercise for you to check that the functions f_j converge uniformly on the entire real line to the function

$$f(x) = \begin{cases} 0 & \text{if} \quad x \le 0 \\ x & \text{if} \quad x > 0 \end{cases}$$

(draw a sketch to help you see this). Notice that each of the functions f_j is continuously differentiable on the entire real line, but f is not differentiable at 0. ∎

It turns out that we must strengthen our convergence hypotheses if we want the limit process to respect differentiation. The basic result is:

Theorem 8.17 *Suppose that a sequence f_j of differentiable functions on an open interval I converges pointwise to a limit function f. Suppose further that the sequence f_j' converges uniformly on I to a limit function g. Then the limit function f is differentiable on I and $f'(x) = g(x)$ for all $x \in I$.*

Proof: Let $\epsilon > 0$. The sequence $\{f_j'\}$ is uniformly Cauchy. Therefore we may choose N so large that $j, k > N$ implies that

$$|f_j'(x) - f_k'(x)| < \frac{\epsilon}{2} \quad \forall x \in I. \tag{8.17.1}$$

Fix a point $c \in I$. Define

$$\mu_j(x) = \frac{f_j(x) - f_j(c)}{x - c}$$

for $x \in I, x \ne c$. It is our intention to apply Theorem 8.11 above to the functions μ_j.

First notice that, for each j, we have

$$\lim_{x \to c} \mu_j(x) = f_j'(c).$$

Thus

$$\lim_{j \to \infty} \lim_{x \to c} \mu_j(x) = \lim_{j \to \infty} f_j'(c) = g(c).$$

That calculates the limits in one order.

On the other hand,

$$\lim_{j \to \infty} \mu_j(x) = \frac{f(x) - f(c)}{x - c} \equiv \mu(x)$$

for $x \in I \setminus \{c\}$. If we can show that this convergence is uniform then Theorem 8.11 applies and we may conclude that

$$\lim_{x \to c} \mu(x) = \lim_{j \to \infty} \lim_{x \to c} \mu_j(x) = \lim_{j \to \infty} f_j'(c) = g(c).$$

But this just says that f is differentiable at c and the derivative equals g. That is the desired result.

To verify the uniform convergence of the μ_j, we apply the Mean Value Theorem to the function $f_j - f_k$. For $x \neq c$ we have

$$
\begin{aligned}
|\mu_j(x) - \mu_k(x)| &= \frac{1}{|x - c|} \cdot |(f_j(x) - f_k(x)) - (f_j(c) - f_k(c))| \\
&= \frac{1}{|x - c|} \cdot |x - c| \cdot |(f_j - f_k)'(\xi)| \\
&= |(f_j - f_k)'(\xi)|
\end{aligned}
$$

for some ξ between x and c. But line (8.17.1) guarantees that the last line does not exceed $\epsilon/2$. That shows that the μ_j converge uniformly and concludes the proof. □

Remark 8.18 A little additional effort shows that we need only assume in the theorem that the functions f_j converge at a single point x_0 in the domain. One of the exercises asks you to prove this assertion.

Notice further that, if we make the additional assumption that each of the functions f_j' is continuous, then the proof of the theorem becomes much easier. For then

$$f_j(x) = f_j(x_0) + \int_{x_0}^x f_j'(t)\, dt$$

by the Fundamental Theorem of Calculus. The hypothesis that the f_j' converge uniformly then implies, by Theorem 8.9, that the integrals converge to

$$\int_{x_0}^x g(t)\, dt.$$

The hypothesis that the functions f_j converge at x_0 then allows us to conclude that the sequence $f_j(x)$ converges for every x to $f(x)$ and

$$f(x) = f(x_0) + \int_{x_0}^x g(t)\, dt.$$

The Fundamental Theorem of Calculus then yields that $f' = g$ as desired.

A Look Back

1. Give a verbal description of what "uniformly Cauchy" means.

2. What is the role of uniform convergence in addressing the issue of commutation of limits?

3. What do we need to assume about a sequence of functions in order to guarantee that their derivatives converge in a natural fashion?

4. What would be a uniform Cauchy condition for series?

Exercises

1. Prove that, if a series of continuous functions converges uniformly, then the sum function is also continuous.

2. Prove Proposition 8.15. Refer to the parallel result in Chapter 3 for some hints.

3. If a sequence of functions f_j on a domain $S \subseteq \mathbb{R}$ has the property that $f_j \to f$ uniformly on S, then does it follow that $(f_j)^2 \to f^2$ uniformly on S? What simple additional hypothesis will make your answer affirmative?

4. Prove the assertion made in Remark 8.18 that Theorem 8.17 is still true if the functions f_j are assumed to converge at just one point (and also that the derivatives f_j' converge uniformly).

5. Assume that f_j are continuous functions on the interval $[0, 1]$. Suppose that $\lim_{j \to \infty} f_j(x)$ exists for each $x \in [0, 1]$ and defines a function f on $[0, 1]$. Further suppose that $f_1 \le f_2 \le \cdots$. Can you conclude that f is continuous?

6. Let $f : \mathbb{R} \to \mathbb{R}$ be a function. We say that f is *piecewise constant* if the real line can be written as the infinite disjoint union of intervals and f is constant on each of those intervals. Now let φ be a continuous function on $[a, b]$. Show that φ can be uniformly approximated by piecewise constant functions.

7. Prove that a sequence of functions is uniformly Cauchy if and only if it converges uniformly.

* 8. Refer to Exercise **6** for terminology. Let f be a piecewise constant function. Show that f is the pointwise limit of polynomials.

* 9. A function is called "piecewise linear" if it is (**i**) continuous and (**ii**) its graph consists of finitely many linear segments. Prove that a continuous function on an interval $[a, b]$ is the uniform limit of a sequence of piecewise linear functions.

* 10. Let f_j be a uniformly convergent sequence of functions on a common domain S. What would be suitable conditions on a function ϕ to guarantee that $\phi \circ f_j$ converges uniformly on S?

* 11. Construct a sequence of continuous functions $f_j(x)$ that has the property that $f_j(q)$ increases monotonically to $+\infty$ for each rational q but such that, at each irrational x, $|f_j(x)| \le 1$ for infinitely many j.

* **12.** Prove that a sequence $\{f_j\}$ of functions converges pointwise if and only if the series

$$f_1 + \sum_{j=2}^{\infty} (f_j - f_{j-1})$$

converges pointwise. Prove the same result for uniform convergence.

8.3 Series of Functions

Preliminary Remarks

Of course a series of functions is understood by studying the sequence of its partial sums. So, in some sense, the theory of sequences of functions and the theory of series of functions are equivalent. But series are useful because they explicitly represent the idea of decomposing an arbitrary function into elemental pieces.

Definition 8.19 The formal expression

$$\sum_{j=1}^{\infty} f_j(x),$$

where the f_j are functions on a common domain S, is called a *series of functions*. For $N = 1, 2, 3, \ldots$ the expression

$$S_N(x) = \sum_{j=1}^{N} f_j(x) = f_1(x) + f_2(x) + \cdots + f_N(x)$$

is called the Nth *partial sum* for the series. In case

$$\lim_{N \to \infty} S_N(x)$$

exists and is finite then we say that the series *converges* at x. Otherwise we say that the series *diverges* at x.

POINT OF CONFUSION 8.20 Notice that the question of convergence of a series of functions, which should be thought of as an *addition process*, reduces to a question about the *sequence* of partial sums. Sometimes, as in the next example, it is convenient to begin the series at some index other than $j = 1$.

EXAMPLE 8.21 Consider the series

$$\sum_{j=0}^{\infty} x^j .$$

This is the geometric series from Proposition 3.21. It converges absolutely for $|x| < 1$ and diverges otherwise.

By the formula for the partial sums of a geometric series,

$$S_N(x) = \frac{1 - x^{N+1}}{1 - x}$$

for $x \neq 1$. When $|x| < 1$ we see that

$$S_N(x) \to \frac{1}{1 - x}.$$

∎

Definition 8.22 Let

$$\sum_{j=1}^{\infty} f_j(x)$$

be a series of functions on a domain S. If the partial sums $S_N(x)$ converge uniformly on S to a limit function $g(x)$, then we say that the series *converges uniformly* on S. We write

$$\sum_{j=1}^{\infty} f_j(x) = g(x)$$

for $x \in S$.

Of course all of our results about uniform convergence of *sequences* of functions translate, via the sequence of partial sums of a series, to results about uniformly convergent series of functions. For example,

(a) If f_j are continuous functions on a domain S and if the series

$$\sum_{j=1}^{\infty} f_j(x)$$

converges uniformly on S to a limit function f, then f is also continuous on S.

(b) If f_j are integrable functions on $[a, b]$ and if

$$\sum_{j=1}^{\infty} f_j(x)$$

converges uniformly on $[a, b]$ to a limit function f, then f is also integrable on $[a, b]$ and

$$\int_a^b f(x)\,dx = \sum_{j=1}^{\infty} \int_a^b f_j(x)\,dx\,.$$

You will be asked to provide details of these assertions, as well as a statement and proof of a result about derivatives of series, in the exercises. Meanwhile we turn to an elegant test for uniform convergence that is due to Weierstrass.

POINT OF CONFUSION 8.23 Notice that, in the last displayed equation, the entity on the left is a number and the entity on the right is a sum of real numbers. This is a statement about the value of an integral.

Theorem 8.24 (The Weierstrass M-Test) *Let $\{f_j\}_{j=1}^{\infty}$ be functions on a common domain S. Assume that each $|f_j|$ is bounded on S by a constant M_j and that*

$$\sum_{j=1}^{\infty} M_j < \infty\,.$$

Then the series

$$\sum_{j=1}^{\infty} f_j \qquad\qquad (8.24.1)$$

converges uniformly and absolutely on the set S.

Proof: By hypothesis, the sequence T_N of partial sums of the series $\sum_{j=1}^{\infty} M_j$ is Cauchy. Given $\epsilon > 0$ there is therefore a number K so large that $q > p > K$ implies that

$$\sum_{j=p+1}^{q} M_j = |T_q - T_p| < \epsilon\,.$$

We may conclude that the partial sums S_N of the original series $\sum f_j$ satisfy, for $q > p > K$,

$$|S_q(x) - S_p(x)| = \left| \sum_{j=p+1}^{q} f_j(x) \right|$$

$$\leq \sum_{j=p+1}^{q} |f_j(x)| \leq \sum_{j=p+1}^{q} M_j < \epsilon.$$

Thus the partial sums $S_N(x)$ of the series (8.24.1) are uniformly Cauchy. The series (8.24.1) therefore converges uniformly. The same estimates apply to the partial sums of the series $\sum_j |f_j|$. Therefore the series also converges absolutely. $\square$

EXAMPLE 8.25 Let us consider the series

$$f(x) = \sum_{j=1}^{\infty} 2^{-j} \sin\left(2^j x\right) .$$

The sine terms oscillate so erratically that it would be difficult to calculate partial sums for this series. However, noting that the jth summand $f_j(x) = 2^{-j} \sin(2^j x)$ is dominated in absolute value by 2^{-j}, we see that the Weierstrass M-Test applies to this series. We conclude that the series converges uniformly and absolutely on the entire real line.

By property (a) of uniformly convergent series of continuous functions that was noted above, we may conclude that the function f defined by our series is continuous. It is also 2π-periodic: $f(x + 2\pi) = f(x)$ for every x since this assertion is true for each summand. Since the continuous function f restricted to the compact interval $[0, 2\pi]$ is uniformly continuous (Theorem 5.30), we may conclude that f is uniformly continuous on the entire real line.

However, it turns out that f is nowhere differentiable. The proof of this assertion follows lines similar to the treatment of nowhere differentiable functions in Theorem 6.7. The details will be covered in an exercise. ∎

A Look Back

1. Say in words what the Weierstrass M-Test says.

2. What does it mean for a series of functions to converge uniformly?

3. What does it mean for a series of functions to converge absolutely?

4. What does it mean for a series of functions to converge pointwise?

5. Why does every question about series reduce to a question about a sequence?

Exercises

1. Formulate and prove a result about the derivative of the sum of a convergent series of differentiable functions.

2. Prove Dini's theorem: If f_j are continuous functions on a compact set K, $f_1(x) \le f_2(x) \le \ldots$ for all $x \in K$, and the f_j converge to a continuous function f on K then in fact the f_j converge *uniformly* to f on K.

3. Use the concept of boundedness of a function to show that the functions $\sin x$ and $\cos x$ cannot be polynomials.

4. Prove that, if p is any polynomial, then there is an N large enough that $e^x > |p(x)|$ for $x > N$. Conclude that the function e^x is not a polynomial.

5. Find a way to prove that $\tan x$ and $\ln x$ are not polynomials.

6. Prove that the series

$$\sum_{j=1}^{\infty} \frac{\sin jx}{j}$$

converges uniformly on compact intervals that do not contain odd multiples of $\pi/2$. (**Hint:** Sum by parts.)

7. Suppose that the sequence $f_j(x)$ on the interval $[0,1]$ satisfies

$$|f_j(s) - f_j(t)| \le |s - t| \tag{$*$}$$

for all $s, t \in [0,1]$. Further assume that the f_j converge pointwise to a limit function f on the interval $[0,1]$. Does the limit function f satisfy $(*)$?

8. Prove a comparison test for uniform convergence of series: if f_j, g_j are functions and $0 \le f_j \le g_j$ and the series $\sum g_j$ converges uniformly then so also does the series $\sum f_j$.

9. Show by giving an example that the converse of the Weierstrass M-Test is false.

10. Prove that if f_j are continuous functions on a domain S and if the series

$$\sum_{j=1}^{\infty} f_j(x)$$

converges uniformly on S to a limit function f, then f is also continuous on S.

11. Prove that if a series $\sum_{j=1}^{\infty} f_j$ of integrable functions on an interval $[a, b]$ is uniformly convergent on $[a, b]$ then the sum function f is integrable and

$$\int_a^b f(x)\, dx = \sum_{j=1}^{\infty} \int_a^b f_j(x)\, dx.$$

* 12. Let $0 < \alpha \le 1$. Prove that the series

$$\sum_{j=1}^{\infty} 2^{-j\alpha} \sin\left(2^j x\right)$$

defines a function f that is nowhere differentiable. To achieve this end, follow the scheme that was used to prove Theorem 6.7: **a)** Fix x; **b)** for h small, choose M such that 2^{-M} is approximately equal to $|h|$; **c)** break the series up into the sum from 1 to $M-1$, the single summand $j = M$, and the sum from $j = M+1$ to ∞. The middle term has very large Newton quotient and the first and last terms are relatively small.

* 13. Give an example of a function that is k-times differentiable but not $(k+1)$-times differentiable at any point.

* 14. Prove that the sequence of functions $f_j(x) = \sin(jx)$ has no subsequence that converges at every x.

8.4   The Weierstrass Approximation Theorem

Preliminary Remarks

One of the most powerful and astonishing theorems of nineteenth century mathematics is the Weierstrass theorem that we study in this section. It tells us that absolutely any continuous function on a closed, bounded interval can be uniformly approximated by a polynomial. This fact has practical significance today, because we cannot program an arbitrary function onto a computer, but we certainly can program a polynomial function.

The name Weierstrass has occurred frequently in this chapter. In fact Karl Weierstrass (1815-1897) revolutionized analysis with his examples and theorems. This section is devoted to one of his most striking results. We introduce it with a motivating discussion.

It is natural to wonder whether the usual functions of calculus—$\sin x$, $\cos x$, and e^x, for instance—are actually polynomials of some very high degree. Since polynomials are so much easier to understand than these transcendental functions, an affirmative answer to this question would certainly simplify mathematics. Of course a moment's thought shows that this wish is impossible: a polynomial of degree k has at most k real roots. Since sine and cosine have infinitely many real roots they cannot be polynomials. A polynomial of degree k has the property that, if it is differentiated enough times (namely, $k + 1$ times), then the derivative is zero. Since this is not the case for e^x, we conclude that e^x cannot be a polynomial. The exercises of the last section discuss other means for distinguishing the familiar transcendental functions of calculus from polynomial functions.

In calculus we learned of a formal procedure, called Taylor series, for associating polynomials with a given function f. In some instances these polynomials form a sequence that converges back to the original function. Of course the method of the Taylor expansion has no hope of working unless f is infinitely differentiable. Even then, it turns out that the Taylor series rarely converges back to the original function. Nevertheless, Taylor's theorem with remainder might cause us to speculate that any reasonable function can be approximated in some fashion by polynomials. In fact the theorem of Weierstrass gives a spectacular affirmation of this speculation:

Theorem 8.26 (The Weierstrass Approximation Theorem) *Let f be a continuous function on an interval $[a, b]$. Then there is a sequence of polynomials $p_j(x)$ with the property that the sequence p_j converges uniformly on $[a, b]$ to f.*

We prove this theorem in detail in the Appendix to this chapter. For now, let us consider some of its consequences. A restatement of the theorem would be that, given a continuous function f on $[a, b]$ and an $\epsilon > 0$, there is a polynomial p such that

$$|f(x) - p(x)| < \epsilon$$

for every $x \in [a, b]$. If one were programming a computer to calculate values of a fairly wild function f, the theorem guarantees that, up to a given degree of accuracy, one could use a polynomial instead (which would in fact be much easier for the computer to handle). Advanced techniques can even tell what degree of polynomial is needed to achieve a given degree of accuracy. The proof that we shall present also suggests how this might be done.

Let f be the Weierstrass Nowhere Differentiable Function. The theorem guarantees that, on any compact interval, f is the uniform limit of polynomials. Thus even the uniform limit of infinitely differentiable functions need not be differentiable—even at one point. This explains why the hypotheses of Theorem 8.17 needed to be so stringent.

A Look Back

1. State verbally what the Weierstrass Approximation Theorem says.

2. Why is Weierstrass's theorem surprising?

3. Is it possible to approximate a continuous function with something other than polynomials?

4. Is there a converse to Weierstrass's theorem?

Exercises

1. Let $\{f_j\}$ be a sequence of continuous functions on the real line. Suppose that the f_j converge uniformly to a function f. Prove that

$$\lim_{j \to \infty} f_j(x + 1/j) = f(x)$$

uniformly on any bounded interval.

Can any of these hypotheses be weakened?

2. Prove that the Weierstrass Approximation Theorem fails if we restrict attention to polynomials of degree less than or equal to 1000.

3. Is the Weierstrass Approximation Theorem true if we restrict ourselves to only using polynomials of even degree?

4. Is the Weierstrass Approximation Theorem true if we restrict ourselves to only using polynomials with coefficients of size not exceeding 1?

5. TRUE or FALSE: If a sequence of polynomials $\{p_j\}$ converges uniformly to 0 on the interval $[-2, 2]$, then the sequence of derivatives $\{p'_j\}$ converges uniformly to 0 on $[-1, 1]$.

6. TRUE or FALSE: If a sequence of polynomials $\{p_j\}$ converges uniformly to 0 on the interval $[-2, 2]$, then the sequence of antiderivatives $\{\int_0^x p_j(t)\,dt\}$ converges uniformly on $[-1, 1]$.

7. Refer to Exercise **9** below for terminology. Prove that the sum of a series of step functions need not be a step function.

* 8. Use the Weierstrass Approximation Theorem and Mathematical Induction to prove that, if f is k times continuously differentiable on an interval $[a, b]$, then there is a sequence of polynomials p_j with the property that

$$p_j \to f$$

uniformly on $[a, b]$,

$$p'_j \to f'$$

uniformly on $[a, b]$,

$$\ldots$$

$$p_j^{(k)} \to f^{(k)}$$

uniformly on $[a, b]$.

* 9. Let $a < b$ be real numbers. Call a function of the form

$$f(x) = \begin{cases} 1 & \text{if} \quad a \le x \le b \\ 0 & \text{if} \quad x < a \text{ or } x > b \end{cases}$$

a *characteristic function* for the interval $[a, b]$. Then a function of the form

$$g(x) = \sum_{j=1}^{k} a_j \cdot f_j(x),$$

with the f_j characteristic functions of intervals $[a_j, b_j]$, is called a *step function*. Prove that any continuous function on an interval $[c, d]$ is the uniform limit of a sequence of step functions. (**Hint:** The proof of this assertion is conceptually easy; do *not* imitate the proof of the Weierstrass Approximation Theorem.)

* 10. If f is a continuous function on the interval $[a, b]$ and if

$$\int_a^b f(x)p(x)\,dx = 0$$

for every polynomial p, then prove that f must be the zero function. (**Hint:** Use Weierstrass's Approximation Theorem.)

* 11. Define a *trigonometric polynomial* to be a function of the form

$$\sum_{j=1}^{k} a_j \cdot \cos jx + \sum_{j=1}^{\ell} b_j \cdot \sin jx.$$

Prove a version of the Weierstrass Approximation Theorem on the interval $[0, 2\pi]$ for 2π-periodic continuous functions and with the phrase "trigonometric polynomial" replacing "polynomial." (**Hint:** Prove that

$$\sum_{\ell=-j}^{j} \left(1 - \frac{|\ell|}{j+1}\right) (\cos \ell t) =$$

$$\frac{1}{j+1} \left(\frac{\sin \frac{j+1}{2} t}{\sin \frac{1}{2} t}\right)^2 .$$

Use these functions as the functions ψ_j in the proof of Weierstrass's theorem.)

APPENDIX: Proof of the Weierstrass Approximation Theorem

We break up the proof of the Weierstrass Approximation Theorem into a sequence of lemmas.

Lemma 8.27 *Let ψ_j be a sequence of continuous functions on the interval $I = [-1, 1]$ with the following properties:*

(i) $\psi_j(x) \geq 0$ *for all x;*

(ii) $\int_{-1}^{1} \psi_j(x)\, dx = 1$ *for each j;*

(iii) *For any $\delta > 0$ we have*

$$\lim_{j \to \infty} \int_{\delta \leq |x| \leq 1} \psi_j(x)\, dx = 0 \,.$$

If f is a continuous function on the real line which is identically zero off the interval $[0, 1]$ then the functions

$$f_j(x) = \int_{-1}^{1} \psi_j(t) f(x - t)\, dt$$

converge uniformly on the interval $[0, 1]$ to $f(x)$.

Proof: By multiplying f by a constant we may assume that $\sup |f| = 1$. Let $\epsilon > 0$. Since f is uniformly continuous on the interval $[0, 1]$ we may choose a $\delta > 0$ such that if $x, t \in I$ $|x - t| < \delta$ then $|f(x) - f(t)| < \epsilon/2$. By property **(iii)** above we may choose an N so large that $j > N$ implies that $|\int_{\delta \leq |t| \leq 1} \psi_j(t)\, dt| < \epsilon/4$. Then, for any $x \in [0, 1]$, we have

$$
\begin{aligned}
|f_j(x) - f(x)| &= \left| \int_{-1}^{1} \psi_j(t) f(x - t)\, dt - f(x) \right| \\
&= \left| \int_{-1}^{1} \psi_j(t) f(x - t)\, dt - \int_{-1}^{1} \psi_j(t) f(x)\, dt \right| .
\end{aligned}
$$

Notice that, in the last line, we have used fact **(ii)** about the functions ψ_j to multiply the term $f(x)$ by 1 in a clever way. Now we may combine the two integrals to find that the last line

$$
= \left| \int_{-1}^{1} (f(x-t) - f(x))\psi_j(t)\, dt \right|
$$

$$
\leq \int_{-\delta}^{\delta} |f(x-t) - f(x)|\psi_j(t)\, dt
$$

$$
+ \int_{\delta \leq |t| \leq 1} |f(x-t) - f(x)|\psi_j(t)\, dt
$$

$$
= A + B.
$$

To estimate term A, we recall that, for $|t| < \delta$, we have $|f(x-t) - f(x)| < \epsilon/2$; hence

$$
A \leq \int_{-\delta}^{\delta} \frac{\epsilon}{2} \psi_j(t)\, dt \leq \frac{\epsilon}{2} \cdot \int_{-1}^{1} \psi_j(t)\, dt = \frac{\epsilon}{2}.
$$

For B we write

$$
B \leq \int_{\delta \leq |t| \leq 1} 2 \cdot \sup |f| \cdot \psi_j(t)\, dt
$$

$$
\leq 2 \cdot \int_{\delta \leq |t| \leq 1} \psi_j(t)\, dt
$$

$$
< 2 \cdot \frac{\epsilon}{4} = \frac{\epsilon}{2},
$$

where in the penultimate line we have used the choice of j. Adding together our estimates for A and B, and noting that these estimates are independent of the choice of x, yields the result. □

Lemma 8.28 Define $\psi_j(t) = k_j \cdot (1-t^2)^j$, where the positive constants k_j are chosen so that $\int_{-1}^{1} \psi_j(t)\, dt = 1$. Then the functions ψ_j satisfy the properties **(i)**–**(iii)** of the last lemma.

Proof: Of course property **(ii)** is true by design. Property **(i)** is obvious. In order to verify property **(iii)**, we need to estimate the size of k_j.

Notice that

$$
\int_{-1}^{1} (1-t^2)^j\, dt = 2 \cdot \int_{0}^{1} (1-t^2)^j\, dt
$$

$$
\geq 2 \cdot \int_{0}^{1/\sqrt{j}} (1-t^2)^j\, dt
$$

$$
\geq 2 \cdot \int_{0}^{1/\sqrt{j}} (1-jt^2)\, dt,
$$

where we have used the binomial theorem and $0 \leq t \leq 1/\sqrt{j}$ to obtain the last inequality. But this last integral is easily evaluated and equals $4/(3\sqrt{j})$. We conclude that

$$\int_{-1}^{1} (1 - t^2)^j \, dt > \frac{1}{\sqrt{j}} \, .$$

As a result, $k_j < \sqrt{j}$.

Now, to verify property **(iii)** of the lemma, we notice that, for $\delta > 0$ fixed and $\delta \leq |t| \leq 1$, it holds that

$$|\psi_j(t)| \leq k_j \cdot (1 - \delta^2)^j \leq \sqrt{j} \cdot (1 - \delta^2)^j$$

and this expression tends to 0 as $j \to \infty$. Thus $\psi_j \to 0$ uniformly on $\{t : \delta \leq |t| \leq 1\}$. It follows that the ψ_j satisfy property **(iii)** of the lemma. $\square$

Proof of the Theorem: We may assume without loss of generality (just by changing coordinates) that f is a continuous function on the interval $[0, 1]$. After adding a linear function (which is a polynomial) to f, we may assume that $f(0) = f(1) = 0$. Thus f may be continued to be a continuous function which is identically zero on $(-\infty, 0]$ and $[1, \infty)$.

Let ψ_j be as in Lemma 8.28 and form f_j as in Lemma 8.27. Then we know that the f_j converge uniformly on $[0, 1]$ to f. Finally,

$$\begin{aligned}
f_j(x) &= \int_{-1}^{1} \psi_j(t) f(x - t) \, dt \\
&= \int_{0}^{1} \psi_j(x - t) f(t) \, dt \\
&= k_j \int_{0}^{1} (1 + (x - t)^2)^j f(t) \, dt \, .
\end{aligned}$$

In the second inequality we performed a simple change of variable.

But multiplying out the expression $(1 + (x - t)^2)^j$ in the integrand then shows that f_j is a polynomial of degree at most $2j$ in x. Thus we have constructed a sequence of polynomials f_j that converges uniformly to the function f on the interval $[0, 1]$. $\square$

Chapter 9

Elementary Transcendental Functions

9.1 Power Series

Preliminary Remarks

When we learn about power series, and especially Taylor's formula, in calculus class we generally come away with the impression that most any function can be expanded in a power series. Unfortunately this is not the case. Functions that have convergent power series expansions are called real-analytic functions, and have many special properties. We shall learn about them in the present section.

A series of the form

$$\sum_{j=0}^{\infty} a_j (x - c)^j$$

is called a *power series* expanded about the point c. Our first task is to determine the nature of the set on which a power series converges.

Proposition 9.1 *Assume that the power series*

$$\sum_{j=0}^{\infty} a_j (x - c)^j$$

converges at the value $x = d$ *with* $d \neq c$. *Let* $r = |d - c|$. *Then the series converges uniformly and absolutely on compact subsets of* $\mathcal{I} = \{x : |x - c| < r\}$.

Proof: We may take the compact subset of $\mathcal{I}$ to be $K = [c - s, c + s]$ for some number $0 < s < r$. For $x \in K$ it then holds that

$$\sum_{j=0}^{\infty} \left| a_j(x - c)^j \right| = \sum_{j=0}^{\infty} \left| a_j(d - c)^j \right| \cdot \left| \frac{x - c}{d - c} \right|^j .$$

In the sum on the right, the first expression in absolute values is bounded by some constant C (by the convergence hypothesis). The quotient in absolute values is majorized by $L = s/r < 1$. The series on the right is thus dominated by

$$\sum_{j=0}^{\infty} C \cdot L^j .$$

This geometric series converges. By the Weierstrass M-Test, the original series converges absolutely and uniformly on K. □

An immediate consequence of the proposition is that the set on which the power series

$$\sum_{j=0}^{\infty} a_j(x - c)^j$$

converges is an interval centered about c. We call this set the *interval of convergence*. The series will converge absolutely and uniformly on compact subsets of the interval of convergence. The *radius* of the interval of convergence (called the *radius of convergence*) is defined to be half its length. Whether convergence holds at the endpoints of the interval will depend on the particular series being studied. Ad hoc methods must be used to check the endpoints. Let us use the notation $\mathcal{P}$ to denote the *open interval of convergence*.

It happens that, if a power series converges at either of the endpoints of its interval of convergence, then the convergence is uniform up to that endpoint. This is a consequence of Abel's partial summation test; details will be explored in the exercises.

On the interval of convergence $\mathcal{P}$, the power series defines a function f. Such a function is said to be *real-analytic*. More precisely, we have

Definition 9.2 A function f, with domain an open set $U \subseteq \mathbb{R}$ and range the real numbers, is called *real analytic* if, for each $c \in U$, the function f may be represented by a convergent power series on an interval of positive radius centered at c:

$$f(x) = \sum_{j=0}^{\infty} a_j(x-c)^j.$$

We need to know both the algebraic and the calculus properties of a real-analytic function: is it continuous? differentiable? How does one add/subtract/multipy/divide two such functions?

Proposition 9.3 *Let*

$$\sum_{j=0}^{\infty} a_j(x-c)^j \quad and \quad \sum_{j=0}^{\infty} b_j(x-c)^j$$

be two power series each having interval of convergence $\mathcal{P}$ centered at c. Let $f(x)$ be the function defined by the first series and $g(x)$ the function defined by the second series. Then, on $\mathcal{P}$, it holds that

(1) $f(x) \pm g(x) = \sum_{j=0}^{\infty}(a_j \pm b_j)(x-c)^j$;

(2) $f(x) \cdot g(x) = \sum_{m=0}^{\infty} \sum_{j+k=m}(a_j \cdot b_k)(x-c)^m.$

Proof: Let

$$A_N = \sum_{j=0}^{N} a_j(x-c)^j \quad and \quad B_N = \sum_{j=0}^{N} b_j(x-c)^j$$

be, respectively, the Nth partial sums of the power series that define f and g. If C_N is the Nth partial sum of the series

$$\sum_{j=0}^{\infty}(a_j \pm b_j)(x-c)^j$$

then

$$f(x) \pm g(x) = \lim_{N \to \infty} A_N \pm \lim_{N \to \infty} B_N = \lim_{N \to \infty}[A_N \pm B_N]$$

$$= \lim_{N \to \infty} C_N = \sum_{j=0}^{\infty}(a_j \pm b_j)(x-c)^j.$$

This proves **(1)**.

For **(2)**, let

$$D_N = \sum_{m=0}^{N} \sum_{j+k=m}(a_j \cdot b_k)(x-c)^m \quad and \quad R_N = \sum_{j=N+1}^{\infty} b_j(x-c)^j.$$

Note that, since the series for g converges, we know that $|R_N(x)| \leq M$ for some positive M.

We have

$$
\begin{aligned}
D_N &= a_0 B_N + a_1(x-c)B_{N-1} + \cdots + a_N(x-c)^N B_0 \\
&= a_0(g(x) - R_N) + a_1(x-c)(g(x) - R_{N-1}) \\
&\quad + \cdots + a_N(x-c)^N(g(x) - R_0) \\
&= g(x) \sum_{j=0}^{N} a_j(x-c)^j \\
&\quad - [a_0 R_N + a_1(x-c)R_{N-1} + \cdots + a_N(x-c)^N R_0].
\end{aligned}
$$

Clearly,

$$
g(x) \sum_{j=0}^{N} a_j(x-c)^j
$$

converges to $g(x)f(x)$ as N approaches ∞. In order to show that $D_N \to g \cdot f$, it will thus suffice to show that

$$
\left| a_0 R_N + a_1(x-c)R_{N-1} + \cdots + a_N(x-c)^N R_0 \right|
$$

converges to 0 as N approaches ∞. Fix x. Now we know that

$$
\sum_{j=0}^{\infty} a_j(x-c)^j
$$

is absolutely convergent so we may set

$$
A = \sum_{j=0}^{\infty} |a_j||x-c|^j .
$$

Also $\sum_{j=0}^{\infty} b_j(x-c)^j$ is convergent. Therefore, given $\epsilon > 0$, we can find N_0 so that $N > N_0$ implies $|R_N| < \epsilon$. Thus we have

$$
\begin{aligned}
&\left| a_0 R_N + a_1(x-c)R_{N-1} + \cdots + a_N(x-c)^N R_0 \right| \\
&\leq \left| a_0 R_N + \cdots + a_{N-N_0}(x-c)^{N-N_0} R_{N_0} \right| \\
&\quad + \left| a_{N-N_0+1}(x-c)^{N-N_0+1} R_{N_0-1} + \cdots + a_N(x-c)^N R_0 \right| \\
&\leq \sup_{M \geq N_0} R_M \cdot \left(\sum_{j=0}^{\infty} |a_j||x-c|^j \right) \\
&\quad + \left| a_{N-N_0+1}(x-c)^{N-N_0+1} R_{N_0-1} \cdots + a_N(x-c)^N R_0 \right| \\
&\leq \epsilon A + \left| a_{N-N_0+1}(x-c)^{N-N_0+1} R_{N_0-1} \cdots + a_N(x-c)^N R_0 \right|.
\end{aligned}
$$

Thus

$$\left| a_0 R_N + a_1 (x - c) R_{N-1} + \cdots + a_N (x - c)^N R_0 \right|$$

$$\leq \quad \epsilon \cdot A + M \cdot \sum_{j=N-N_0+1}^{N} |a_j| |x - c|^j .$$

Since the series defining A converges, we find on letting $N \to \infty$ that

$$\limsup_{N \to \infty} \left| a_0 R_N + a_1 (x - c) R_{N-1} + \cdots + a_N (x - c)^N R_0 \right| \leq \epsilon \cdot A .$$

Since $\epsilon > 0$ was arbitrary, we may conclude that

$$\lim_{N \to \infty} \left| a_0 R_N + a_1 (x - c) R_{N-1} + \cdots + a_N (x - c)^N R_0 \right| = 0 . \qquad \square$$

POINT OF CONFUSION 9.4 Observe that the form of the product of two power series provides some motivation for the form that the product of numerical series took in Theorem 3.50.

POINT OF CONFUSION 9.5 If we are going to manipulate or combine two power series, then we must assume that they are both defined for the same values of x. We may assume that they have the same interval of convergence, or we may work on the *intersection* of their intervals of convergence.

Next we turn to division of real-analytic functions. If f and g are real analytic functions, both defined on an open interval I, and if g does not vanish on I, then we would like f/g to be a well-defined real-analytic function (it certainly is a well-defined *function*) and we would like to be able to calculate its power series expansion by formal long division. This is what the next result tells us.

Proposition 9.6 Let f and g be real-analytic functions, both of which are defined on an open interval I. Assume that g does not vanish on I. Then the function

$$h(x) = \frac{f(x)}{g(x)}$$

is real-analytic on I. Moreover, if I is centered at the point c and if

$$f(x) = \sum_{j=0}^{\infty} a_j (x - c)^j \quad \text{and} \quad g(x) = \sum_{j=0}^{\infty} b_j (x - c)^j ,$$

then the power series expansion of h about c may be obtained by formal long division of the latter series into the former. That is, the zeroeth coefficient c_0 of h is

$$c_0 = a_0/b_0 \,,$$

the order one coefficient c_1 is

$$c_1 = \frac{1}{b_0} \left(a_1 - \frac{a_0 b_1}{b_0} \right) \,,$$

etc.

Proof: If we can show that the power series

$$\sum_{j=0}^{\infty} c_j (x - c)^j$$

converges on I then the result on multiplication of series in Proposition 9.2 yields this new result. There is no loss of generality in assuming that $c = 0$. Assume for the moment that $b_1 \neq 0$.

Notice that one may check inductively that, for $j \geq 1$,

$$c_j = \frac{1}{b_0} (a_j - b_1 \cdot c_{j-1}) \,. \tag{9.6.1}$$

Without loss of generality, we may scale the a_j and the b_j terms and assume that the radius of I is $1 + \epsilon$, some $\epsilon > 0$. Then we see from the last displayed formula that

$$|c_j| \leq C \cdot (|a_j| + |c_{j-1}|) \,,$$

where $C = \max\{|1/b_0|, |b_1/b_0|\}$. It follows inductively that

$$|c_j| \leq C' \cdot (1 + |a_j| + |a_{j-1}| + \cdots + |a_0|) \,,$$

Since the radius of I exceeds 1, $\sum |a_j| < \infty$ and we see that the $|c_j|$ are bounded. Hence the power series with coefficients c_j has radius of convergence 1.

In case $b_1 = 0$ then the role of b_1 is played by the first nonvanishing $b_m, m > 1$. Then a new version of formula (9.6.1) is obtained and the argument proceeds as before. $\qquad \square$

In practice it is often useful to calculate f/g by expanding g in a "geometric series." To illustrate this idea, we assume for simplicity that f and g

are real-analytic in a neighborhood of 0. Then

$$\frac{f(x)}{g(x)} = f(x) \cdot \frac{1}{g(x)}$$

$$= f(x) \cdot \frac{1}{b_0 + b_1 x + \cdots}$$

$$= f(x) \cdot \frac{1}{b_0} \cdot \frac{1}{1 + (b_1/b_0)x + \cdots}.$$

Now we use the fact that, for β small,

$$\frac{1}{1-\beta} = 1 + \beta + \beta^2 + \cdots.$$

Setting $\beta = -(b_1/b_0)x - (b_2/b_0)x^2 - \cdots$ and substituting the resulting expansion into our expression for $f(x)/g(x)$ then yields a formula that can be multiplied out to give a power series expansion for $f(x)/g(x)$. We explore this technique in the exercises.

A Look Back

1. If f is an infinitely differentiable function on the interval $(-1,1)$, then what is the jth coefficient of its Taylor expansion?

2. What is a necessary and sufficient condition for the Taylor expansion of a function to converge?

3. What does the zero set of a real-analytic function look like?

4. Why is the power series expansion of a real-analytic function unique?

Exercises

1. Prove that the composition of two real-analytic functions, when the composition makes sense, is also real-analytic.

2. Prove that
$$\sin^2 x + \cos^2 x = 1$$
directly from the power series expansions.

3. Let $f(x) = \sum_{j=0}^{\infty} a_j x^j$ be a power series convergent on the interval $(-r, r)$ and let Z denote those points in the interval where f vanishes. Prove that if Z has an accumulation point in the interval then $f \equiv 0$. (**Hint:** If a is the accumulation point, expand f in a power series about a. What is the first nonvanishing term in that expansion?)

* 4. Prove the assertion from the text that, if a power series converges at an endpoint of the interval of convergence, then the convergence is uniform up to that endpoint.

5. Use the technique described at the end of this section to calculate the first three terms of the power series expansion of $\sin x / e^x$ about the origin.

6. Provide the details of the method for dividing real-analytic functions that is described at the end of the section.

7. Use the technique described at the end of this section to calculate the first three terms of the power series expansion of $\ln x / \sin(\pi x / 2)$ about $c = 1$.

8. Prove that $\cos 2x = \cos^2 x - \sin^2 x$ directly from the power series expansions.

9. Show that the solution of the differential equation $y' + y = x$ will be real-analytic.

10. Prove that $\sin 2x = 2 \sin x \cos x$ directly from the power series expansions.

11. Show that the inverse of a (suitable) real-analytic function is real-analytic.

* 12. Verify that the function

$$f(x) = \begin{cases} 0 & \text{if} \quad x = 0 \\ e^{-1/x^2} & \text{if} \quad x \neq 0 \end{cases}$$

is infinitely differentiable on all of $\mathbb{R}$ and that $f^{(k)}(0) = 0$ for every k. However, f is certainly not real-analytic.

9.2   More on Power Series: Convergence Issues

Preliminary Remarks

In this section we learn Hadamard's elegant formula for the radius of convergence of a power series. And we learn more about the Taylor expansion. As previously noted, any smooth function has a (formal) Taylor expansion. The issue is whether it converges, and whether it converges back to the original function.

We now introduce the *Hadamard formula* for the radius of convergence of a power series.

Lemma 9.7 (Hadamard) *For the power series*

$$\sum_{j=0}^{\infty} a_j (x - c)^j ,$$

define A and ρ by

$$A = \limsup_{n \to \infty} |a_n|^{1/n} ,$$

$$\rho = \begin{cases} 0 & \text{if } A = \infty, \\ 1/A & \text{if } 0 < A < \infty, \\ \infty & \text{if } A = 0, \end{cases}$$

then ρ is the radius of convergence of the power series about c.

Proof: Observing that

$$\limsup_{n \to \infty} |a_n(x - c)^n|^{1/n} = A|x - c|,$$

we see that the lemma is an immediate consequence of the Root Test. $\quad\Box$

Corollary 9.8 *The power series*

$$\sum_{j=0}^{\infty} a_j(x - c)^j$$

has radius of convergence ρ if and only if, when $0 < R < \rho$, there exists a constant $0 < C = C_R$ such that

$$|a_n| \le \frac{C}{R^n}.$$

POINT OF CONFUSION 9.9 The single most important attribute of a power series is its radius of convergence. It turns out that the radius of convergence is best understood in the context of the complex numbers. We cannot say much about that point in the present book, but some exercises will touch on the issue.

From the power series

$$\sum_{j=0}^{\infty} a_j(x - c)^j$$

it is natural to create the *derived series*

$$\sum_{j=1}^{\infty} j a_j(x - c)^{j-1}$$

using term-by-term differentiation.

Proposition 9.10 *The radius of convergence of the derived series is the same as the radius of convergence of the original power series.*

Proof: We observe that

$$\limsup_{n\to\infty} |na_n|^{1/n} = \lim_{n\to\infty} n^{-1/n} \limsup_{n\to\infty} |na_n|^{1/n}$$
$$= \limsup_{n\to\infty} |a_n|^{1/n}.$$

So the result follows from the Hadamard formula. □

Proposition 9.11 *Let f be a real-analytic function defined on an open interval I. Then f is continuous and has continuous, real-analytic derivatives of all orders. In fact the derivatives of f are obtained by differentiating its series representation term by term.*

Proof: Since, for each $c \in I$, the function f may be represented by a convergent power series with positive radius of convergence, we see that, in a sufficiently small open interval about each $c \in I$, the function f is the uniform limit of a sequence of continuous functions: the partial sums of the power series representing f. It follows that f is continuous at c. Since the radius of convergence of the derived series is the same as that of the original series, it also follows that the derivatives of the partial sums converge uniformly on an open interval about c to a continuous function. It then follows from Theorem 8.17 that f is differentiable and its derivative is the function defined by the derived series. By induction, f has continuous derivatives of all orders at c. □

We can now show that a real-analytic function has a unique power series representation at any point.

Corollary 9.12 *If the function f is represented by a convergent power series on an interval of positive radius centered at c,*

$$f(x) = \sum_{j=0}^{\infty} a_j (x - c)^j ,$$

then the coefficients of the power series are related to the derivatives of the function by

$$a_n = \frac{f^{(n)}(c)}{n!} .$$

Proof: This follows readily by differentiating both sides of the above equation n times, as we may by the proposition, and evaluating at $x = c$. □

Finally, we note that integration of power series is as well-behaved as differentiation.

Proposition 9.13 *The power series*

$$\sum_{j=0}^{\infty} a_j (x - c)^j$$

and the series

$$\sum_{j=0}^{\infty} \frac{a_j}{j+1} (x - c)^{j+1}$$

obtained from term-by-term integration have the same radius of convergence, and the function F defined by

$$F(x) = \sum_{j=0}^{\infty} \frac{a_j}{j+1} (x - c)^{j+1}$$

on the common interval of convergence satisfies

$$F'(x) = \sum_{j=0}^{\infty} a_j (x - c)^j = f(x).$$

Proof: The proof is left to the exercises. □

We conclude this section with a consideration of Taylor series:

Theorem 9.14 (Taylor's Expansion) *For k a nonnegative integer, let f be a $k+1$ times continuously differentiable function on an open interval $I = (a - \epsilon, a + \epsilon)$. Then, for $x \in I$,*

$$f(x) = \sum_{j=0}^{k} f^{(j)}(a) \frac{(x - a)^j}{j!} + R_{k,a}(x),$$

where

$$R_{k,a}(x) = \int_a^x f^{(k+1)}(t) \frac{(x - t)^k}{k!} \, dt.$$

Proof: We apply integration by parts to the Fundamental Theorem of Calculus to obtain

$$
\begin{aligned}
f(x) &= f(a) + \int_a^x f'(t) \, dt \\
&= f(a) + \left(f'(t) \frac{(t - x)}{1!} \right) \Big|_a^x - \int_a^x f''(t) \frac{(t - x)}{1!} \, dt \\
&= f(a) + f'(a) \frac{(x - a)}{1!} + \int_a^x f''(t) \frac{x - t}{1!} \, dt.
\end{aligned}
$$

Notice that, when we performed the integration by parts, we used $t - x$ as an antiderivative for dt. This is of course legitimate, as a glance at the integration by parts theorem reveals. We have proved the theorem for the case $k = 1$. The result for higher values of k is obtained inductively by repeated applications of integration by parts.   □

Taylor's theorem allows us to associate with any infinitely differentiable function a formal expansion of the form

$$\sum_{j=0}^{\infty} a_j (x - a)^j .$$

However, there is no guarantee that this series will converge. Even if it does converge, it may not converge back to $f(x)$. An important example to keep in mind is the function

$$h(x) = \begin{cases} 0 & \text{if} \quad x = 0 \\ e^{-1/x^2} & \text{if} \quad x \neq 0. \end{cases}$$

This function is infinitely differentiable at every point of the real line (including the point 0—use l'Hôpital's Rule). However, all of its derivatives at $x = 0$ are equal to zero (this matter will be treated in the exercises). Therefore the formal Taylor series expansion of h about $a = 0$ is

$$\sum_{j=0}^{\infty} 0 \cdot (x - 0)^j = 0 .$$

We see that the formal Taylor series expansion for h converges to the zero function at every x, but not to the original function h itself.

In fact the theorem tells us that the Taylor expansion of a function f converges to f at a point x if and only if $R_{k,a}(x) \to 0$. In the exercises we shall explore the following more quantitative assertion:

An infinitely differentiable function f on an interval I has Taylor series expansion about $a \in I$ that converges back to f on a neighborhood J of a if and only if there are positive constants C, R such that, for every $x \in J$ and every k, it holds that

$$\left| f^{(k)}(x) \right| \leq C \cdot \frac{k!}{R^k} .$$

The function h considered above should not be thought of as an isolated exception. For instance, we know from calculus that the function $f(x) = \sin x$ has Taylor expansion that converges to f at every x. But then, for ϵ small, the function $g_\epsilon(x) = f(x) + \epsilon \cdot h(x)$ has Taylor series that does *not* converge back to $g_\epsilon(x)$ for $x \neq 0$. Similar examples may be generated by using other real analytic functions in place of sine.

A Look Back

1. What estimate on the derivatives of a smooth function will imply that it is real-analytic?

2. How does one prove Taylor's expansion?

3. How are the radius of convergence of a power series and the radius of convergence of its derived series related?

4. How are the radius of convergence of a power series and the radius of convergence of its integrated series related?

Exercises

1. Let f be an infinitely differentiable function on an interval I. If $a \in I$ and there are positive constants C, R such that, for every x in a neighborhood of a and every k, it holds that

$$\left| f^{(k)}(x) \right| \le C \cdot \frac{k!}{R^k},$$

then prove that the Taylor series of f about a converges to $f(x)$. (**Hint:** estimate the error term.) What is the radius of convergence?

2. Let f be an infinitely differentiable function on an open interval I centered at a. Assume that the Taylor expansion of f about a converges to f at every point of I. Prove that there are constants C, R and a (possibly smaller) interval J centered at a such that, for each $x \in J$, it holds that

$$\left| f^{(k)}(x) \right| \le C \cdot \frac{k!}{R^k}.$$

3. Give examples of power series, centered at 0, on the interval $(-1, 1)$, which **(a)** converge only on $(-1, 1)$, **(b)** converge only on $[-1, 1)$, **(c)** converge only on $(-1, 1]$, **(d)** converge only on $[-1, 1]$.

* 4. Prove that, if a function on an interval I has derivatives of all orders which are positive at every point of I, then f is real-analytic on I.

5. Prove Proposition 9.13.

6. The real-analytic function $1/(1 + x^2)$ is well defined on the entire real line. Yet its power series about 0 only converges on an interval of radius 1. Explain why. [**Hint:** Think in terms of the complex numbers.]

7. How does Exercise **6** differ for the function $1/(1 - x^2)$?

* 8. Show that the function

$$h(x) = \begin{cases} 0 & \text{if} \quad x = 0 \\ e^{-1/x^2} & \text{if} \quad x \ne 0. \end{cases}$$

is infinitely differentiable (you will need to take special care at the origin). What is its Taylor expansion at 0?

* **9.** The function defined by a power series may extend continuously to an endpoint of the interval of convergence without the series converging at that endpoint. Give an example.

* **10.** Assume that a power series converges at one of the endpoints of its interval of convergence. Use summation by parts to prove that the function defined by the power series is continuous on the half-closed interval including that endpoint.

11. Prove Proposition 9.12.

* **12.** If $\{a_j\}_{j=0}^\infty$ is any sequence of real numbers then there is an infinitely differentiable function whose power series expansion about 0 is $\sum_j a_j x^j$. Explain why. [This is a theorem of E. Borel.]

9.3   The Exponential and Trigonometric Functions

Preliminary Remarks

In the present section we give rigorous definitions of the exponential and trigonometric functions, and derive some of their most basic properties. We also take a look at the important constant π.

We begin by defining the exponential function:

Definition 9.15 The power series

$$\sum_{j=0}^{\infty} \frac{x^j}{j!}$$

converges, by the Ratio Test, for every real value of x. The function defined thereby is called the *exponential function* and is written $\exp(x)$.

POINT OF CONFUSION 9.16 The only other straightforward way to define the exponential function is that it is the solution of the differential equation $y' = y$. You can, if you wish, substitute the power series in Definition 9.15 into this differential equation to see that it is satisfied.

Proposition 9.17 *The function* $\exp(x)$ *satisfies*

$$\exp(a + b) = \exp(a) \cdot \exp(b)$$

for any complex numbers a and b.

Proof: We write the righthand side as

$$\left(\sum_{j=0}^{\infty}\frac{a^j}{j!}\right)\cdot\left(\sum_{j=0}^{\infty}\frac{b^j}{j!}\right).$$

Now convergent power series may be multiplied term by term. We find that the last line equals

$$\sum_{j=0}^{\infty}\left(\sum_{\ell=0}^{j}\frac{a^{(j-\ell)}}{(j-\ell)!}\cdot\frac{b^{\ell}}{\ell!}\right). \tag{9.16.1}$$

However, the inner sum on the right side of this equation may be written as

$$\frac{1}{j!}\sum_{\ell=0}^{j}\frac{j!}{\ell!(j-\ell)!}a^{j-\ell}b^{\ell}=\frac{1}{j!}(a+b)^j.$$

It follows that line (9.16.1) equals $\exp(a+b)$. □

We set $e=\exp(1)$. This is consistent with our earlier treatment of the number e in Section 3.4. The proposition tells us that, for any positive integer k, we have

$$e^k=e\cdot e\cdots e=\exp(1)\cdot\exp(1)\cdots\exp(1)=\exp(k).$$

If m is another positive integer then

$$(\exp(k/m))^m=\exp(k)=e^k,$$

whence

$$\exp(k/m)=e^{k/m}.$$

We may extend this formula to *negative* rational exponents by using the fact that $\exp(a)\cdot\exp(-a)=1$. Thus, for any rational number q,

$$\exp(q)=e^q.$$

Now note that the function exp is increasing and continuous. It follows (this fact is treated in the exercises) that if we set, for any $r\in\mathbb{R}$,

$$e^r=\sup\{e^q:q\in\mathbb{Q}\text{ and }q<r\}$$

(this is a *definition* of the expression e^r), then $e^x=\exp(x)$ for every real x. [You may find it useful to review the discussion of exponentiation in Sections 2.4, 3.4; the presentation here parallels those treatments.] We will adhere to custom and write e^x instead of $\exp(x)$.

Proposition 9.18 *The exponential function e^x, for $x \in \mathbb{R}$, satisfies*

(a) $e^x > 0$ *for all* x;

(b) $e^0 = 1$;

(c) $(e^x)' = e^x$;

(d) e^x *is strictly increasing;*

(e) *the graph of e^x is asymptotic to the negative x-axis*

(f) *for each integer $N > 0$ there is a number c_N such that $e^x > c_N \cdot x^N$ when $x > 0$.*

Proof: The first three statements are obvious from the power series expansion for the exponential function.

If $s < t$ then the Mean Value Theorem tells us that there is a number ξ between s and t such that

$$e^t - e^s = (t - s) \cdot e^\xi > 0;$$

hence the exponential function is strictly increasing.

By inspecting the power series we see that $e^x > 1 + x$ hence e^x increases to $+\infty$. Since $e^x \cdot e^{-x} = 1$ we conclude that e^{-x} tends to 0 as $x \to +\infty$. Thus the graph of the exponential function is asymptotic to the negative x-axis.

Finally, by inspecting the power series for e^x, we see that the last assertion is true with $c_N = 1/N!$.   □

Now we turn to the trigonometric functions. The definition of the trigonometric functions that is found in calculus texts is unsatisfactory because it relies too heavily on a picture and because the continual need to subtract off superfluous multiples of 2π is clumsy. We have nevertheless used the trigonometric functions in earlier chapters to illustrate various concepts. It is time now to give a rigorous definition of the trigonometric functions that is independent of these earlier considerations.

Definition 9.19 The power series

$$\sum_{j=0}^{\infty} (-1)^j \frac{x^{2j+1}}{(2j+1)!}$$

converges at every point of the real line (by the Ratio Test). The function that it defines is called the *sine* function and is usually written $\sin x$.

The power series

$$\sum_{j=0}^{\infty} (-1)^j \frac{x^{2j}}{(2j)!}$$

converges at every point of the real line (by the Ratio Test). The function that it defines is called the *cosine* function and is usually written $\cos x$.

You may recall that the power series that we use to define the sine and cosine functions are precisely the Taylor series expansions for the functions sine and cosine that were derived in your calculus text. But now we *begin* with the power series and must derive the properties of sine and cosine that we need *from these series*.

In fact the most convenient way to achieve this goal is to proceed by way of the exponential function. [The point here is mainly one of convenience. It can be verified by direct manipulation of the power series that $\sin^2 x + \cos^2 x = 1$ and so forth but the algebra is extremely unpleasant.] The formula in the next proposition is usually credited to Euler.

POINT OF CONFUSION 9.20 For Euler's result, we will use the complex numbers, in particular the special number i. Of course you know that $i^2 = -1$. More generally, you know that

$$(a + ib) \cdot (c + id) = (ac - bd) + i(bc + ad).$$

Proposition 9.21 *The exponential function and the functions sine and cosine are related by the formula (for x and y real and $i^2 = -1$)*

$$\exp(x + iy) = e^x \cdot (\cos y + i \sin y) .$$

Proof: We shall verify the case $x = 0$ and leave the general case for the reader.

Thus we are to prove that

$$e^{iy} = \cos y + i \sin y . \tag{9.21.1}$$

Writing out the power series for the exponential, we find that the lefthand side of (9.21.1) is

$$\sum_{j=0}^{\infty} \frac{(iy)^j}{j!}$$

and this equals

$$\left[1 - \frac{y^2}{2!} + \frac{y^4}{4!} - + \cdots\right] + i\left[\frac{y}{1!} - \frac{y^3}{3!} + \frac{y^5}{5!} - + \cdots\right].$$

Of course the two series on the right are the familiar power series for cosine and sine. Thus

$$e^{iy} = \cos y + i \sin y,$$

as desired.   □

In what follows, we think of the formula (9.21.1) as *defining* what we mean by e^{iy}. As a result,

$$e^{x+iy} = e^x \cdot e^{iy} = e^x \cdot (\cos y + i \sin y).$$

Notice that $e^{-iy} = \cos(-y) + i \sin(-y) = \cos y - i \sin y$ (we know that the sine function is odd and the cosine function even from their power series expansions). Then formula (9.21.1) tells us that

$$\cos y = \frac{e^{iy} + e^{-iy}}{2}$$

and

$$\sin y = \frac{e^{iy} - e^{-iy}}{2i}.$$

Now we may prove:

Proposition 9.22 *For every real x it holds that*

$$\sin^2 x + \cos^2 x = 1.$$

Proof: Simply substitute into the left side the formulas for the sine and cosine functions which were displayed before the proposition and simplify.   □

We list several other properties of the sine and cosine functions that may be proved by similar methods. The proofs are requested of you in the exercises.

Proposition 9.23 *The functions sine and cosine have the following properties:*

(a) $\sin(s + t) = \sin s \cos t + \cos s \sin t$;

(b) $\cos(s + t) = \cos s \cos t - \sin s \sin t$;

(c) $\cos(2s) = \cos^2 s - \sin^2 s$;

(d) $\sin(2s) = 2 \sin s \cos s$;

(e) $\sin(-s) = -\sin s$;

(f) $\cos(-s) = \cos s$;

(g) $\sin'(s) = \cos s$;

(h) $\cos'(s) = -\sin s$.

One important task to be performed in a course on the foundations of analysis is to define the number π and establish its basic properties. In a course on Euclidean geometry, the constant π is defined to be the ratio of the circumference of a circle to its diameter. Such a definition is not useful for our purposes (however, it *is* consistent with the definition about to be given here).

Observe that $\cos 0$ is the real part of e^{i0} which is 1. Thus if we set

$$\alpha = \inf\{x > 0 : \cos x = 0\}$$

then $\alpha > 0$ and, by the continuity of the cosine function, $\cos \alpha = 0$. We define $\pi = 2\alpha$.

Applying Proposition 9.22 to the number α yields that $\sin \alpha = \pm 1$. Since α is the *first* zero of cosine on the right half line, the cosine function must be positive on $(0, \alpha)$. But cosine is the derivative of sine. Thus the sine function is *increasing* on $(0, \alpha)$. Since $\sin 0$ is the imaginary part of e^{i0} which is 0, we conclude that $\sin \alpha > 0$ hence that $\sin \alpha = +1$.

Now we may apply parts **(c)** and **(d)** of Proposition 9.23 with $s = \alpha$ to conclude that $\sin \pi = 0$ and $\cos \pi = -1$. A similar calculation with $s = \pi$ shows that $\sin 2\pi = 0$ and $\cos 2\pi = 1$. Next we may use parts **(a)** and **(b)** of Proposition 9.23 to calculate that $\sin(x + 2\pi) = \sin x$ and $\cos(x + 2\pi) = \cos x$ for all x. In other words, the sine and cosine functions are 2π–periodic.

The business of calculating a decimal expansion for π would take us far afield. One approach would be to utilize the already-noted fact that the sine function is strictly increasing on the interval $[0, \pi/2]$ hence its inverse function

$$\mathrm{Sin}^{-1} : [0, 1] \to [0, \pi/2]$$

is well defined. Then one can determine (see Chapter 6) that

$$\left(\mathrm{Sin}^{-1}\right)'(x) = \frac{1}{\sqrt{1 - x^2}} \, .$$

By the Fundamental Theorem of Calculus,

$$\frac{\pi}{2} = \mathrm{Sin}^{-1}(1) = \int_0^1 \frac{1}{\sqrt{1 - x^2}} \, dx \, .$$

By approximating the integral by its Riemann sums, one obtains an approximation to $\pi/2$ and hence to π itself. This approach will be explored in more detail in the exercises.

Let us for now observe that

$$
\begin{aligned}
\cos 2 &= 1 - \frac{2^2}{2!} + \frac{2^4}{4!} - \frac{2^6}{6!} + - \cdots \\
&= 1 - 2 + \frac{16}{24} - \frac{64}{720} + \ldots.
\end{aligned}
$$

Since the series defining $\cos 2$ is an alternating series with terms that strictly decrease to zero in magnitude, we may conclude (following reasoning from Chapter 4) that the last line is less than the sum of the first three terms:

$$
\cos 2 < -1 + \frac{2}{3} < 0.
$$

It follows that $\alpha = \pi/2 < 2$ hence $\pi < 4$. A similar calculation of $\cos(3/2)$ would allow us to conclude that $\pi > 3$.

A Look Back

1. How do we define the exponential function?

2. How do we define $\sin x$ and $\cos x$?

3. How is the exponential function related to sine and cosine?

4. How do we know that the exponential function is not a polynomial?

Exercises

1. Provide the details of the assertion preceding Proposition 9.18 to the effect that if we define, for any real r,

$$
e^r = \sup\{e^q : q \in \mathbb{Q} \text{ and } q < r\},
$$

then $e^x = \exp(x)$ for every real x.

2. Prove the equality $\left(\mathrm{Sin}^{-1}\right)'(x) = 1/\sqrt{1 - x^2}$.

3. Find a formula for $\cos 4x$ directly from the power series expansions.

4. Find a formula for $\sin(x + \pi/2)$ directly from the power series expansions.

5. Use one of the methods described at the end of Section 9.3 to calculate π to two decimal places.

6. Prove Proposition 9.22.

7. Prove parts (a), (b), (c), (d) of Proposition 9.23.

8. Prove that the trigonometric polynomials, that is to say, the functions of the form

$$p(x) = \sum_{j=-N}^{N} a_j \cos jx + b_j \sin jx,$$

are dense in the continuous functions on $[0, 2\pi]$ in the uniform topology.

9. Prove the general case of Proposition 9.21.

10. Prove parts (e), (f), (g), (h) of Proposition 9.23.

11. Find a formula for $\tan^4 x$ in terms of $\sin 2x$, $\sin 4x$, $\cos 2x$, and $\cos 4x$.

12. Refer to Exercise 1 for notation and terminology. Prove that the exponential function defined there is one-to-one and onto from the real line to the half-line.

13. Define $\log x = \int_1^x 1/t \, dt$. Using this definition of logarithm, derive the basic properties of the log function that are discussed in the text.

9.4 Logarithms and Powers of Real Numbers

Preliminary Remarks

The logarithm function is of interest because it is the inverse of the exponential function, but also because it is used to define entropy in physics. The logarithm is useful in understanding exponential functions in general. We study these matters in the present section.

Since the exponential function $\exp(x) = e^x$ is positive and strictly increasing, it is a one-to-one function from $\mathbb{R}$ to $(0, \infty)$. Thus it has a well-defined inverse function that we call the *natural logarithm*. We write this function as $\ln x$.

Proposition 9.24 *The natural logarithm function has the following properties:*

(a) $(\ln x)' = 1/x$;

(b) $\ln x$ *is strictly increasing;*

(c) $\ln 1 = 0$;

(d) $\ln e = 1$;

(e) *the graph of the natural logarithm function is asymptotic to the negative y-axis;*

(f) $\ln(s \cdot t) = \ln s + \ln t$;

(g) $\ln(s/t) = \ln s - \ln t$.

Proof: These follow immediately from corresponding properties of the exponential function. For example, to verify part **(f)**, set $s = e^\sigma$ and $t = e^\tau$. Then

$$
\begin{aligned}
\ln(s \cdot t) &= \ln(e^\sigma \cdot e^\tau) \\
&= \ln(e^{\sigma+\tau}) \\
&= \sigma + \tau \\
&= \ln s + \ln t.
\end{aligned}
$$

The other parts of the proposition are proved similarly. $\qquad\Box$

Proposition 9.25 *If a and b are positive real numbers then*

$$
a^b = e^{b \cdot \ln a} \, .
$$

Proof: When b is an integer then the formula may be verified directly from the definition of logarithm. For $b = m/n$ a rational number the formula follows by our usual trick of passing to nth roots. For arbitrary b we use a limiting argument as in our discussions of exponentials in Sections 2.3 and 9.3. $\qquad\Box$

Remark 9.26 We have discussed several different approaches to the exponentiation process. We proved the existence of nth roots, $n \in \mathbb{N}$, as an illustration of the completeness of the real numbers (by taking the supremum of a certain set). We treated rational exponents by composing the usual arithmetic process of taking mth powers with the process of taking nth roots. Then, in Sections 2.4 and 9.3, we passed to arbitrary powers by way of a limiting process.

Proposition 9.25 gives us a unified and direct way to treat all exponentials at once. This unified approach will prove (see the next proposition) to be particularly advantageous when we wish to perform calculus operations on exponential functions.

Proposition 9.27 *Fix $a > 0$. The function $f(x) = a^x$ has the following properties:*

(a) $(a^x)' = a^x \cdot \ln a$;

(b) $f(0) = 1$;

(c) if $0 < a < 1$ then f is decreasing and the graph of f is asymptotic to the positive x-axis;

(d) if $1 < a$ then f is increasing and the graph of f is asymptotic to the negative x-axis.

Proof: These properties follow immediately from corresponding properties of the function exp. □

The logarithm function arises, among other places, in the context of probability and in the study of entropy. The reason is that the logarithm function is uniquely determined by the way that it interacts with the operation of multiplication:

Theorem 9.28 Let $\phi(x)$ be a continuously differentiable function with domain the positive reals and which satisfies the identity

$$\phi(s \cdot t) = \phi(s) + \phi(t) \tag{9.28.1}$$

for all positive s and t. Then there is a constant $C > 0$ such that

$$\phi(x) = C \cdot \ln x$$

for all x.

Proof: Differentiate the equation (9.28.1) with respect to s to obtain

$$t \cdot \phi'(s \cdot t) = \phi'(s) \,.$$

Now fix s and set $t = 1/s$ to conclude that

$$\phi'(1) \cdot \frac{1}{s} = \phi'(s) \,.$$

We take the constant C to be $\phi'(1)$ and apply Proposition 9.24(**a**) to conclude that $\phi(s) = C \cdot \ln s + D$ for some constant D. But ϕ cannot satisfy (9.28.1) unless $D = 0$, so the theorem is proved. □

Observe that the *natural logarithm function* is then the unique continuously differentiable function that satisfies the condition (9.28.1) and whose derivative at 1 equals 1. That is the reason that the natural logarithm function (rather than the common logarithm, or logarithm to the base ten) is singled out as the focus of our considerations in this section.

A Look Back

1. What is the definition of the natural logarithm function?
2. What is the definition of the common logarithm function?
3. How can we use the logarithm to define a^b for arbitrary positive a and any real b?
4. What is the derivative of the logarithm function?

Exercises

1. Prove Proposition 9.25 by following the suggested line of reasoning.
2. Prove Proposition 9.24, except for part **(f)**.
3. Prove that condition (9.28.1) implies that $\phi(1) = 0$. Assume that ϕ is differentiable at $x = 1$ but make no other hypothesis about the smoothness of ϕ. Prove that condition (9.28.1) then implies that ϕ is differentiable at every $x > 0$.
4. Provide the details of the proof of Proposition 9.27.
5. Calculate
$$\lim_{j \to \infty} \frac{j^{j/2}}{j!}.$$
6. Give three distinct reasons why the natural logarithm function is not a polynomial.
7. At infinity, any nontrivial polynomial function dominates the natural logarithm function. Explain what this means, and prove it.
8. At infinity, any exponential function with base greater than 1 dominates any polynomial. Explain this statement and prove it.
9. Prove that
$$a^{b^c} = a^{bc}.$$

* 10. The *Lambert W function* is defined implicitly by the equation
$$z = W(z) \cdot e^{W(z)}.$$
   It is a fact that any elementary transcendental function (sine, cosine, logarithm, exponential) may be expressed (with an elementary formula) in terms of the W function. Prove that this is so for the exponential function and the sine function.

* 11. Refer to Exercise **10**. Show that the Lambert W function is real-analytic on its domain.

* 12. Show that the hypothesis of Theorem 9.28 may be replaced with $f \in \text{Lip}_\alpha([0, 2\pi])$, some $\alpha > 0$.

* 13. Prove Euler's formula relating the exponential to sine and cosine *not* by using power series, but rather by using differential equations.

Appendix I: Elementary Number Systems

Section A1.1. The Natural Numbers

Mathematics deals with a variety of number systems. The simplest number system in real analysis is $\mathbb{N}$, the *natural numbers*. As we have already noted, this is just the set of positive integers $\{1, 2, 3, \ldots\}$. In a rigorous course of logic, the set $\mathbb{N}$ is constructed from the axioms of set theory. However, in this book we shall assume that you are familiar with the positive integers and their elementary properties.

The principal properties of $\mathbb{N}$ are as follows:

1. 1 is a natural number.

2. If x is a natural number then there is another natural number $\widehat{x}$ which is called the *successor* of x.

3. $1 \neq \widehat{x}$ for every natural number x.

4. If $\widehat{x} = \widehat{y}$ then $x = y$.

5. *(Principle of Induction)* If $\mathcal{P}$ is a property and if

   (a) 1 has the property $\mathcal{P}$;

   (b) whenever a natural number x has the property $\mathcal{P}$ it follows that $\widehat{x}$ also has the property $\mathcal{P}$;

   then all natural numbers have the property $\mathcal{P}$.

These rules, or *axioms*, are known as the Peano Axioms for the natural numbers (named after Giuseppe Peano (1858-1932) who developed them). We take it for granted that the usual set of positive integers satisfies these rules. Certainly 1 is in that set. Each positive integer has a "successor"—after 1

217

comes 2 and after 2 comes 3 and so forth. The number 1 is not the successor of any other positive integer. Two positive integers with the same successor must be the same. The last axiom is more challenging but makes good sense: if some property $\mathcal{P}(n)$ holds for $n = 1$ and if whenever it holds for n then it also holds for $n + 1$, then we may conclude that $\mathcal{P}$ holds for all positive integers.

We will spend the remainder of this section exploring Axiom **5**, the Principle of Induction.

Example A1.1

Let us prove that for each positive integer n it holds that

$$1 + 2 + \cdots + n = \frac{n \cdot (n + 1)}{2}.$$

We denote this equation by $\mathcal{P}(n)$, and follow the scheme of the Principle of Induction.

First, $\mathcal{P}(1)$ is true since then both the left and the right side of the equation equal 1. Now assume that $\mathcal{P}(n)$ is true for some natural number n. Our job is to show that it follows that $\mathcal{P}(n + 1)$ is true.

Since $\mathcal{P}(n)$ is true, we know that

$$1 + 2 + \cdots + n = \frac{n \cdot (n + 1)}{2}.$$

Let us add the quantity $n + 1$ to both sides. Thus

$$1 + 2 + \cdots + n + (n + 1) = \frac{n \cdot (n + 1)}{2} + (n + 1).$$

The right side of this new equality simplifies and we obtain

$$1 + 2 + \cdots + (n + 1) = \frac{(n + 1) \cdot ((n + 1) + 1)}{2}.$$

But this is just $\mathcal{P}(n + 1) = \mathcal{P}(\widehat{n})$. *We have assumed* $\mathcal{P}(n)$ *and have proved* $\mathcal{P}(\widehat{n})$, just as the Principle of Induction requires.

Thus we may conclude that property $\mathcal{P}$ holds for all positive integers, as desired. ∎

The formula that we derived in Example A1.1 was probably known to the ancient Greeks. However, a celebrated anecdote credits Karl Friedrich Gauss (1777-1855) with discovering the formula when he was nine years old. Gauss went on to become (along with Isaac Newton and Archimedes) one of the three greatest mathematicians of all time.

The formula from Example A1.1 gives a neat way to add up the integers from 1 to n for any n, without doing any work. Any time that we discover a new mathematical fact, there are generally several others hidden within it. The next example illustrates this point.

Example A1.2

The sum of the first m positive even integers is $m \cdot (m + 1)$. To see this note that the sum in question is

$$2 + 4 + 6 + \cdots + 2m = 2(1 + 2 + 3 + \cdots + m).$$

But, by the first example, the sum in parentheses on the right is equal to $m \cdot (m + 1)/2$. It follows that

$$2 + 4 + 6 + \cdots + 2m = 2 \cdot \frac{m \cdot (m + 1)}{2} = m \cdot (m + 1). \qquad \square$$

The second example could also be performed by induction (without using the result of the first example). ∎

Example A1.3

Now we will use induction incorrectly to prove a statement that is completely preposterous:

All horses are the same color.

There are finitely many horses in existence, so it is convenient for us to prove the slightly more technical statement

Any collection of k horses consists of horses
which are all the same color.

Our statement $\mathcal{P}(k)$ is this last displayed statement.

Now $\mathcal{P}(1)$ is true: *one horse is the same color.* (Note: this is not a joke, and the error has not occurred yet.)

Suppose next that $\mathcal{P}(k)$ is true: we assume that any collection of k horses has the same color. Now consider a collection of $\widehat{k} = k + 1$ horses. Remove one horse from that collection. By our hypothesis, the remaining k horses have the same color.

Now replace the horse that we removed and remove a different horse. Again, the remaining k horses have the same color.

We keep repeating this process: remove each of the $k + 1$ horses one by one and conclude that the remaining k horses have the same color. Therefore every horse in the collection is the same color as every other. So all $k + 1$ horses have the same color. The statement $\mathcal{P}(k+1)$ is thus proved (assuming the truth of $\mathcal{P}(k)$) and the induction is complete.

Where is our error? It is nothing deep—just an oversight. The argument we have given is wrong when $\widehat{k} = k + 1 = 2$. For remove one horse from a set of two and the remaining (*one*) horse is the same color. Now replace the removed horse and remove the other horse. The remaining (*one*) horse is the same color. *So what?* We cannot conclude that the two horses are colored the same. Thus the induction breaks down at the outset; the reasoning is incorrect. ∎

Proposition A1.4

Let a and b be real numbers and n a natural number. Then

$$
\begin{aligned}
(a + b)^n &= a^n + \frac{n}{1}a^{n-1}b + \frac{n(n-1)}{2 \cdot 1}a^{n-2}b^2 \\
&\quad + \frac{n(n-1)(n-2)}{3 \cdot 2 \cdot 1}a^{n-3}b^3 \\
&\quad + \cdots + \frac{n(n-1)\cdots 2}{(n-1)(n-2)\cdots 2 \cdot 1}ab^{n-1} + b^n.
\end{aligned}
$$

Proof: The case $n = 1$ being obvious, proceed by induction. □

Example A1.5

The expression

$$
\frac{n(n-1)\cdots(n-k+1)}{k(k-1)\cdots 1}
$$

is often called the *kth binomial coefficient* and is denoted by the symbol

$$
\binom{n}{k}.
$$

Using the notation $m! = m \cdot (m-1) \cdot (m-2) \cdots 2 \cdot 1$, for m a natural number, we may write the kth binomial coefficient as

$$
\binom{n}{k} = \frac{n!}{(n-k)! \cdot k!}.
$$

∎

Section A1.2. The Integers

Now we will apply the notion of an equivalence class to *construct* the integers (both positive and negative). There is an important point of knowledge to be noted here. For the sake of having a reasonable place to begin our work, we took the natural numbers $\mathbb{N} = \{1, 2, 3, \ldots\}$ as given. Since the natural numbers have been used for thousands of years to keep track of objects for barter, this is a plausible thing to do. Even people who know no mathematics accept the positive integers. However, the number zero and the negative numbers are a different matter. It was not until the fifteenth century that the concepts of zero and negative numbers started to take hold—for they do not correspond to explicit collections of objects (five fingers or ten shoes) but rather to *concepts* (zero books is the lack of books; minus 4 pens means that we owe someone four pens). After some practice we get used to negative numbers, but explaining in words what they mean is always a bit clumsy.

It is much more satisfying, from the point of view of logic, to *construct* the integers (including the negative whole numbers and zero) from what we already have, that is, from the natural numbers. We proceed as follows. Let $A = \mathbb{N} \times \mathbb{N}$, the set of ordered pairs of natural numbers. We define a relation (see Appendix II, Section A2.6) $\mathcal{R}$ on A and A as follows:

$$(a, b) \text{ is related to } (a', b') \text{ if } a + b' = a' + b$$

See also Appendix II, Section A2.6 for the concept of equivalence relation.

Theorem A1.6

The relation $\mathcal{R}$ is an equivalence relation.

Proof: That (a, b) is related to (a, b) follows from the trivial identity $a + b = a + b$. Hence $\mathcal{R}$ is reflexive. Second, if (a, b) is related to (a', b') then $a + b' = a' + b$ hence $a' + b = a + b'$ (just reverse the equality) hence (a', b') is related to (a, b). So $\mathcal{R}$ is symmetric.

Finally, if (a, b) is related to (a', b') and (a', b') is related to (a'', b'') then we have

$$a + b' = a' + b \qquad \text{and} \qquad a' + b'' = a'' + b'.$$

Adding these equations gives

$$(a + b') + (a' + b'') = (a' + b) + (a'' + b').$$

Cancelling a' and b' from each side finally yields

$$a + b'' = a'' + b.$$

Thus (a, b) is related to (a'', b''). Therefore $\mathcal{R}$ is transitive. We conclude that $\mathcal{R}$ is an equivalence relation. $\square$

Now our job is to understand the equivalence classes which are induced by $\mathcal{R}$. [We will ultimately call this number system the integers $\mathbb{Z}$.] Let $(a, b) \in A$ and let $[(a, b)]$ be the corresponding equivalence class. If $b > a$ then we will denote this equivalence class by the integer $b - a$. For instance, the equivalence class $[(2, 7)]$ will be denoted by 5. Notice that if $a < b$ and $(a', b') \in [(a, b)]$ then $a + b' = a' + b$ hence $b' - a' = b - a$. Therefore the integer symbol that we choose to represent our equivalence class is *independent of which element of the equivalence class is used to compute it.*

If $(a, b) \in A$ and $b = a$ then we let the symbol 0 denote the equivalence class $[(a, b)]$. Notice that if (a', b') is any other element of $[(a, b)]$ then it must be that $a + b' = a' + b$ hence $b' = a'$; therefore this definition is unambiguous.

If $(a, b) \in A$ and $a > b$ then we will denote the equivalence class $[(a, b)]$ by the symbol $-(a - b)$. For instance, we will denote the equivalence class $[(7, 5)]$ by the symbol -2. Once again, if $a > b$ and if (a', b') is related to (a, b) then the equation $a + b' = a' + b$ guarantees that our choice of symbol to represent $[(a, b)]$ is unambiguous.

Thus we have given our equivalence classes names, and these names *look just like* the names that we usually give to integers: there are positive integers, and negative ones, and zero. But we want to see that these objects *behave* like integers. (As you read on, use the intuitive, non-rigorous mnemonic that the equivalence class $[(a, b)]$ stands for the integer $b - a$.)

First, do these new objects that we have constructed *add* correctly? Well, let $X = [(a, b)]$ and $Y = [(c, d)]$ be two equivalence classes. *Define* their sum to be $X + Y = [(a + c, b + d)]$. We must check that this is unambiguous. If $(\tilde{a}, \tilde{b})$ is related to (a, b) and $(\tilde{c}, \tilde{d})$ is related to (c, d) then of course we know that

$$a + \tilde{b} = \tilde{a} + b$$

and

$$c + \tilde{d} = \tilde{c} + d.$$

Adding these two equations gives

$$(a + c) + (\tilde{b} + \tilde{d}) = (\tilde{a} + \tilde{c}) + (b + d)$$

hence $(a + c, b + d)$ is related to $(\tilde{a} + \tilde{c}, \tilde{b} + \tilde{d})$. Thus, adding two of our equivalence classes gives another equivalence class, as it should.

Example A1.7

To add 5 and 3 we first note that 5 is the equivalence class $[(2,7)]$ and 3 is the equivalence class $[(2,5)]$. We add them componentwise and find that the sum is $[(2+2, 7+5)] = [(4,12)]$. Which equivalence class is this answer? Looking back at our prescription for giving names to the equivalence classes, we see that this is the equivalence class that we called $12 - 4$ or 8. So we have rediscovered the fact that $5 + 3 = 8$. Check for yourself that, if we were to choose a different representative for 5—say $(6,11)$—and a different representative for 3—say $(24,27)$—then the same answer would result.

Now let us add 4 and -9. The first of these is the equivalence class $[(3,7)]$ and the second is the equivalence class $[(13,4)]$. The sum is therefore $[(16,11)]$, and this is the equivalence class that we call $-(16 - 11)$ or -5. That is the answer that we would expect when we add 4 to -9.

Next, we add -12 and -5. Previous experience causes us to expect the answer to be -17. Now -12 is the equivalence class $[(19,7)]$ and -5 is the equivalence class $[(7,2)]$. The sum is $[(26,9)]$, which is the equivalence class that we call -17.

Finally, we can see in practice that our method of addition is unambiguous. Let us redo the second example using $[(6,10)]$ as the equivalence class represented by 4 and $[(15,6)]$ as the equivalence class represented by -9. Then the sum is $[(21,16)]$, and this is still the equivalence class -5, as it should be. ∎

The assertion that the result of calculating a sum—no matter which representatives we choose for the equivalence classes—will give only one answer is called the "fact that addition is *well defined*." In order for our definitions to make sense, it is essential that we check this property of well-definedness.

Remark A1.8

What is the point of this section? Everyone knows about negative numbers, so why go through this abstract construction? The reason is that, until one sees this construction, negative numbers are just imaginary objects—placeholders if you will—which are a useful notation but which do not exist. Now they *do* exist. They are a collection of equivalence classes of pairs of natural numbers. This collection is equipped with certain arithmetic operations, such as addition, subtraction, and multiplication. We now discuss these last two.

If $x = [(a,b)]$ and $y = [(c,d)]$ are integers, we define their *difference* to be the equivalence class $[(a+d, b+c)]$; we denote this difference by $x - y$.

Remark A1.9

We calculate $8 - 14$. Now $8 = [(1, 9)]$ and $14 = [(3, 17)]$. Therefore

$$8 - 14 = [(1 + 17, 9 + 3)] = [(18, 12)] = -6,$$

as expected.

As a second example, we compute $(-4) - (-8)$. Now

$$-4 - (-8) = [(6, 2)] - [(13, 5)] = [(6 + 5, 2 + 13)] = [(11, 15)] = 4.$$

Remark A1.10

When we first learn that $(-4) - (-8) = (-4) + 8 = 4$, the explanation is a bit mysterious: why is "minus a minus equal to a plus"? Now there is no longer any mystery: this property follows *from our construction* of the number system $\mathbb{Z}$. ∎

Finally, we turn to multiplication. If $x = [(a, b)]$ and $y = [(c, d)]$ are integers then we define their product by the formula

$$x \cdot y = [(a \cdot d + b \cdot c, a \cdot c + b \cdot d)].$$

This definition may be a surprise. Why did we not define $x \cdot y$ to be $[(a \cdot c, b \cdot d)]$? There are several reasons: first of all, the latter definition would give the wrong answer; moreover, it is not unambiguous (different representatives of x and y would give a different answer). If you recall that we think of $[(a, b)]$ as representing $b - a$ and $[(c, d)]$ as representing $d - c$ then the product should be the equivalence class that represents $(b - a) \cdot (d - c)$. That is the motivation behind our definition.

We proceed now to an example.

Example A1.11

We compute the product of -3 and -6. Now

$$(-3) \cdot (-6) = [(5, 2)] \cdot [(9, 3)] = [(5 \cdot 3 + 2 \cdot 9, 5 \cdot 9 + 2 \cdot 3)] = [(33, 51)] = 18,$$

which is the expected answer.

As a second example, we multiply -5 and 12. We have

$$-5 \cdot 12 = [(7, 2)] \cdot [(1, 13)] = [(7 \cdot 13 + 2 \cdot 1, 7 \cdot 1 + 2 \cdot 13)] = [(93, 33)] = -60.$$

Finally, we show that 0 times any integer A equals zero. Let $A = [(a, b)]$. Then

$$0 \cdot A = [(1, 1)] \cdot [(a, b)] = [(1 \cdot b + 1 \cdot a, 1 \cdot a + 1 \cdot b)]$$
$$= [(a + b, a + b)]$$
$$= 0. \qquad \blacksquare$$

Remark A1.12

Notice that one of the pleasant byproducts of our construction of the integers is that we no longer have to give artificial explanations for why the product of two negative numbers is a positive number or why the product of a negative number and a positive number is negative. These properties instead follow automatically from our construction.

Of course we will not discuss division for integers; in general division of one integer by another makes no sense *in the universe of the integers.*[1]

In the rest of this book we will follow the standard mathematical custom of denoting the set of all integers by the symbol $\mathbb{Z}$. We will write the integers not as equivalence classes, but in the usual way as $\cdots - 3, -2, -1, 0, 1, 2, 3, \ldots$. The equivalence classes are a device that we used to *construct* the integers. Now that we have the integers in hand, we may as well write them in the simple, familiar fashion.

In an exhaustive treatment of the construction of $\mathbb{Z}$, we would prove that addition and multiplication are commutative and associative, prove the distributive law, and so forth. But the purpose of this section is to demonstrate modes of logical thought rather than to be thorough.

Section A1.3. The Rational Numbers

In this section we use the integers, together with a construction using equivalence classes, to build the rational numbers. Let A be the set $\mathbb{Z} \times (\mathbb{Z} \setminus \{0\})$. Here the symbol $\setminus$ stands for "subtraction of sets": $\mathbb{Z} \setminus \{0\}$ denotes the set of all elements of $\mathbb{Z}$ *except* 0. In other words, A is the set of ordered pairs (a, b) of integers subject to the condition that $b \neq 0$. [*Think, intuitively and non-rigorously, of this ordered pair as "representing" the fraction a/b.*] We

---

[1] Here the word "universe" is meant to mean the collection of all objects on which we are focusing our attention.

definitely want it to be the case that certain ordered pairs represent the same number. For instance,

$$\text{The number } \tfrac{1}{2} \text{ should be the same number as } \tfrac{3}{6}.$$

This example motivates our equivalence relation. Declare (a, b) to be related to (a', b') if $a \cdot b' = a' \cdot b$. [*Here we are thinking, intuitively and non-rigorously, that the fraction a/b should equal the fraction a'/b' precisely when $a \cdot b' = a' \cdot b$.*]

Is this an equivalence relation? Obviously the pair (a, b) is related to itself, since $a \cdot b = a \cdot b$. Also the relation is symmetric: if (a, b) and (a', b') are pairs and $a \cdot b' = a' \cdot b$ then $a' \cdot b = a \cdot b'$. Finally, if (a, b) is related to (a', b') and (a', b') is related to (a'', b'') then we have both

$$a \cdot b' = a' \cdot b \quad \text{and} \quad a' \cdot b'' = a'' \cdot b'.$$

Multiplying the left sides of these two equations together and the right sides together gives

$$(a \cdot b') \cdot (a' \cdot b'') = (a' \cdot b) \cdot (a'' \cdot b').$$

If $a' = 0$ then it follows immediately from the two equations preceding the last that both a and a'' must be zero. So the three pairs $(a, b), (a', b')$, and (a'', b'') are equivalent and there is nothing to prove. So we may assume that $a' \neq 0$. We know *a priori* that $b' \neq 0$; therefore we may cancel common terms in the last equation to obtain

$$a \cdot b'' = b \cdot a''.$$

Thus (a, b) is related to (a'', b''), and our relation is transitive.

The resulting collection of equivalence classes will be called the set of *rational numbers*, and we shall denote this set with the symbol $\mathbb{Q}$.

Example A1.13

The equivalence class $[(4, 12)]$ in the rational numbers contains all of the pairs $(4, 12), (1, 3), (-2, -6)$. (Of course it contains infinitely many other pairs as well.) This equivalence class represents the fraction $4/12$, which we sometimes also write as $1/3$ or $-2/(-6)$. ∎

If $[(a, b)]$ and $[(c, d)]$ are rational numbers then we define their *product* to be the rational number

$$[(a \cdot c, b \cdot d)].$$

This is well defined, for if (a, b) is related to $(\tilde{a}, \tilde{b})$ and (c, d) is related to $(\tilde{c}, \tilde{d})$ then we have the equations

$$a \cdot \widetilde{b} = \widetilde{a} \cdot b \quad \text{and} \quad c \cdot \widetilde{d} = \widetilde{c} \cdot d.$$

Multiplying together the left sides and the right sides we obtain

$$(a \cdot \widetilde{b}) \cdot (c \cdot \widetilde{d}) = (\widetilde{a} \cdot b) \cdot (\widetilde{c} \cdot d).$$

Rearranging, we have

$$(a \cdot c) \cdot (\widetilde{b} \cdot \widetilde{d}) = (\widetilde{a} \cdot \widetilde{c}) \cdot (b \cdot d).$$

But this says that the product of $[(a, b)]$ and $[(c, d)]$ is related to the product of $[(\widetilde{a}, \widetilde{b})]$ and $[(\widetilde{c}, \widetilde{d})]$. So multiplication is unambiguous (i.e., well defined).

Example A1.14

The product of the two rational numbers $[(3, 8)]$ and $[(-2, 5)]$ is

$$[(3 \cdot (-2), 8 \cdot 5)] = [(-6, 40)] = [(-3, 20)].$$

This is what we expect: the product of $3/8$ and $-2/5$ is $-3/20$.   ∎

If $q = [(a, b)]$ and $r = [(c, d)]$ are rational numbers and if r is not zero (that is, $[(c, d)]$ is not the equivalence class zero—in other words, $c \neq 0$) then we define the quotient q/r to be the equivalence class

$$[(ad, bc)].$$

We leave it to you to check that this operation is well defined.

Example A1.15

The quotient of the rational number $[(4, 7)]$ by the rational number $[(3, -2)]$ is, by definition, the rational number

$$[(4 \cdot (-2), 7 \cdot 3)] = [(-8, 21)].$$

This is what we expect: the quotient of $4/7$ by $-3/2$ is $-8/(21)$.   ∎

How should we add two rational numbers? We could try declaring $[(a, b)] + [(c, d)]$ to be $[(a + c, b + d)]$, but this will not work (think about the way that we usually add fractions). Instead we define

$$[(a, b)] + [(c, d)] = [(a \cdot d + c \cdot b, b \cdot d)].$$

We turn to an example.

Example A1.16

The sum of the rational numbers $[(3, -14)]$ and $[(9, 4)]$ is given by

$$[(3 \cdot 4 + 9 \cdot (-14), (-14) \cdot 4)] = [(-114, -56)] = [(57, 28)].$$

This coincides with the usual way that we add fractions:

$$-\frac{3}{14} + \frac{9}{4} = \frac{57}{28}.$$ ∎

Notice that the equivalence class $[(0, 1)]$ is the rational number that we usually denote by 0. It is the additive identity, for if $[(a, b)]$ is another rational number then

$$[(0, 1)] + [(a, b)] = [(0 \cdot b + a \cdot 1, 1 \cdot b)] = [(a, b)].$$

A similar argument shows that $[(0, 1)]$ times any rational number $[(a, b)]$ gives $[(0, b)]$ or 0.

Of course the concept of subtraction is really just a special case of addition (that is $x - y$ is the same thing as $x + (-y)$). So we shall say nothing further about subtraction.

In practice we will write rational numbers in the traditional fashion:

$$\frac{2}{5}, \quad \frac{-19}{3}, \quad \frac{22}{2}, \quad \frac{24}{4}, \quad \ldots$$

In mathematics it is generally not wise to write rational numbers in mixed form, such as $2\frac{3}{5}$, because the juxtaposition of two numbers could easily be mistaken for multiplication. Instead we would write this quantity as the improper fraction $13/5$.

Definition A1.17

A set S is called a *field* if it is equipped with a binary operation (usually called addition and denoted "$+$") and a second binary operation (called multiplication and denoted "$\cdot$") such that the following axioms are satisfied:

A1. S is closed under addition: if $x, y \in S$ then $x + y \in S$.

A2. Addition is commutative: if $x, y \in S$ then $x + y = y + x$.

A3. Addition is associative: if $x, y, z \in S$ then $x + (y + z) = (x + y) + z$.

A4. There exists an element, called 0, in S which is an additive identity: if $x \in S$ then $0 + x = x$.

A5. Each element of S has an additive inverse: if $x \in S$ then there is an element $-x \in S$ such that $x + (-x) = 0$.

M1. S is closed under multiplication: if $x, y \in S$ then $x \cdot y \in S$.

M2. Multiplication is commutative: if $x, y \in S$ then $x \cdot y = y \cdot x$.

M3. Multiplication is associative: if $x, y, z \in S$ then $x \cdot (y \cdot z) = (x \cdot y) \cdot z$.

M4. There exists an element, called 1, which is a multiplicative identity: if $x \in S$ then $x \cdot 1 = x$.

M5. Each nonzero element of S has a multiplicative inverse: if $0 \neq x \in S$ then there is an element $x^{-1} \in S$ such that $x \cdot (x^{-1}) = 1$. The element x^{-1} is sometimes denoted $1/x$.

D1. Multiplication distributes over addition: if $x, y, z \in S$ then

$$x \cdot (y + z) = x \cdot y + x \cdot z .$$

Eleven axioms is a lot to digest all at once, but in fact these are all familiar properties of addition and multiplication of rational numbers that we use every day: the set $\mathbb{Q}$, with the usual notions of addition and multiplication, forms a field. The integers, by contrast, do not: nonzero elements of $\mathbb{Z}$ (except 1 and -1) do not have multiplicative inverses *in the integers*.

Let us now consider some consequence of the field axioms.

Theorem A1.18

Any field has the following properties:

(1) If $z + x = z + y$ then $x = y$.

(2) If $x + z = 0$ then $z = -x$ *(the additive inverse is unique)*.

(3) $-(-y) = y$.

(4) If $y \neq 0$ and $y \cdot x = y \cdot z$ then $x = z$.

(5) If $y \neq 0$ and $y \cdot z = 1$ then $z = y^{-1}$ *(the multiplicative inverse is unique)*.

(6) $\left(x^{-1}\right)^{-1} = x$.

(7) $0 \cdot x = 0$.

(8) If $x \cdot y = 0$ then either $x = 0$ or $y = 0$.

(9) $(-x) \cdot y = -(x \cdot y) = x \cdot (-y)$.

(10) $(-x) \cdot (-y) = x \cdot y$.

Proof: These are all familiar properties of the rationals, but now we are considering them for an arbitrary field. We prove just a few to illustrate the logic.

To prove **(1)** we write

$$z + x = z + y \Rightarrow (-z) + (z + x) = (-z) + (z + y)$$

and now Axiom **A3** yields that this implies

$$((-z) + z) + x = ((-z) + z) + y \,.$$

Next, Axiom **A5** yields that

$$0 + x = 0 + y$$

and hence, by Axiom **A4**,

$$x = y \,.$$

To prove **(7)**, we observe that

$$0 \cdot x = (0 + 0) \cdot x \,,$$

the righthand side of which by Axiom **M2** equals

$$x \cdot (0 + 0).$$

By Axiom **D1** the last expression equals

$$x \cdot 0 + x \cdot 0 \,,$$

which by Axiom **M2** equals $0 \cdot x + 0 \cdot x$. Thus we have derived the equation

$$0 \cdot x = 0 \cdot x + 0 \cdot x \,.$$

Axioms **A4** and **A2** let us rewrite the left side as

$$0 \cdot x + 0 = 0 \cdot x + 0 \cdot x \,.$$

Finally, part **(1)** of the present theorem (which we have already proved) yields that

$$0 = 0 \cdot x \,,$$

which is the desired result.

To prove (8), we suppose that $x \neq 0$. In this case x has a multiplicative inverse x^{-1} and we multiply both sides of our equation by this element:

$$x^{-1} \cdot (x \cdot y) = x^{-1} \cdot 0.$$

By Axiom **M3**, the left side can be rewritten and we have

$$(x \cdot x^{-1}) \cdot y = x^{-1} \cdot 0.$$

Next, we rewrite the right side using Axiom **M2**:

$$(x \cdot x^{-1}) \cdot y = 0 \cdot x^{-1}.$$

Now Axiom **M5** allows us to simplify the left side:

$$1 \cdot y = 0 \cdot x^{-1}.$$

We further simplify the left side using Axiom **M4** and the right side using Part (7) of the present theorem (which we just proved) to obtain:

$$y = 0.$$

Thus we see that if $x \neq 0$ then $y = 0$. But this is logically equivalent with $x = 0$ or $y = 0$, as we wished to prove. [If you have forgotten why these statements are logically equivalent, write a truth table (for which concept see Appendix II).] □

Definition A1.19

Let A be a set. We shall say that A is *ordered* if there is a relation $\mathcal{R}$ on A and A satisfying the following properties:

1. If $a \in A$ and $b \in A$ then one and only one of the following holds: $(a, b) \in \mathcal{R}$ or $(b, a) \in \mathcal{R}$ or $a = b$.

2. If a, b, c are elements of A and $(a, b) \in \mathcal{R}$ and $(b, c) \in \mathcal{R}$ then $(a, c) \in \mathcal{R}$.

We call the relation $\mathcal{R}$ an *order* on A.

Rather than write an ordering relation as $(a, b) \in \mathcal{R}$ it is usually more convenient to write it as $a < b$. The notation $b > a$ means the same thing as $a < b$.

Example A1.20

The integers $\mathbb{Z}$ form an ordered set with the usual ordering $<$. We can make this ordering precise by saying that $x < y$ if $y - x$ is a positive integer. For instance,

$$6 < 8 \text{ because } 8 - 6 = 2 > 0.$$

Likewise,

$$-5 < -1 \text{ because } -1 - (-5) = 4 > 0.$$

Observe that the same ordering works on the rational numbers.   ∎

If A is an ordered set and a, b are elements then we often write $a \leq b$ to mean that *either $a = b$ or $a < b$.*

When a field has an ordering which is compatible with the field operations then a richer structure results:

Definition A1.21

A field F is called an *ordered field* if F has an ordering $<$ that satisfies the following addition properties:

(1) If $x, y, z \in F$ and $y < z$ then $x + y < x + z$.

(2) If $x, y \in F, x > 0$, and $y > 0$ then $x \cdot y > 0$.

Again, these are familiar properties of the rational numbers: $\mathbb{Q}$ forms an ordered field. But there are many other ordered fields as well (for instance, the real numbers $\mathbb{R}$ form an ordered field).

Theorem A1.22
Any ordered field has the following properties:

(1) *If $x > 0$ and $z < y$ then $x \cdot z < x \cdot y$.*

(2) *If $x < 0$ and $z < y$ then $x \cdot z > x \cdot y$.*

(3) *If $x > 0$ then $-x < 0$. If $x < 0$ then $-x > 0$.*

(4) *If $0 < y < x$ then $0 < 1/x < 1/y$.*

(5) *If $x \neq 0$ then $x^2 > 0$.*

(6) *If $0 < x < y$ then $x^2 < y^2$.*

Proof: Again we prove just a few of these statements.

To prove **(1)**, observe that the property **(1)** of ordered fields together with our hypothesis implies that

$$(-z) + z < (-z) + y \,.$$

Thus, using **(A2)**, we see that $y - z > 0$. Since $x > 0$, property **(2)** of ordered fields gives

$$x \cdot (y - z) > 0 \,.$$

Finally,

$$x \cdot y = x \cdot [(y - z) + z] = x \cdot (y - z) + x \cdot z > 0 + x \cdot z$$

(by property **(1)** again). In conclusion,

$$x \cdot y > x \cdot z \,.$$

To prove **(3)**, begin with the equation

$$0 = -x + x \,.$$

Since $x > 0$, the right side is greater than $-x$. Thus $0 > -x$ as claimed. The proof of the other statement of **(3)** is similar.

To prove **(5)**, we consider two cases. If $x > 0$ then $x^2 \equiv x \cdot x$ is positive by property **(2)** of ordered fields. If $x < 0$ then $-x > 0$ (by part **(3)** of the present theorem, which we just proved) hence $(-x) \cdot (-x) > 0$. But part **(10)** of the last theorem guarantees that $(-x) \cdot (-x) = x \cdot x$ hence we see that $x \cdot x > 0$. □

We conclude this Appendix by recording an inadequacy of the field of rational numbers; this will serve in part as motivation for learning about the real numbers in Section 1.1:

Theorem A1.23
There is no positive rational number q such that $q^2 = q \cdot q = 2$.

Proof: Seeking a contradiction, suppose that there is such a q. Write q in lowest terms as

$$q = \frac{a}{b},$$

with a and b greater than zero. This means that the numbers a and b have no common divisors except 1. The equation $q^2 = 2$ can then be written as

$$a^2 = 2 \cdot b^2 \,.$$

Since 2 divides the right side of this last equation, it follows that 2 divides the left side. But 2 can divide a^2 only if 2 divides a (because 2 is prime). We write $a = 2 \cdot \alpha$ for some positive integer α. But then the last equation becomes

$$4 \cdot \alpha^2 = 2 \cdot b^2 \,.$$

Simplifying yields that

$$2 \cdot \alpha^2 = b^2 \,.$$

Since 2 divides the left side, we conclude that 2 must divide the right side. But 2 can divide b^2 only if 2 divides b.

This is our contradiction: we have argued that 2 divides a *and* that 2 divides b. But a and b were assumed to *have no common divisors*. We conclude that the rational number q cannot exist.   □

In fact it turns out that a positive integer can be the square of a rational number if and only if it is the square of a positive integer. This assertion is a special case of a more general phenomenon in number theory known as Gauss's lemma.

Appendix II: Logic and Set Theory

Everyday language is imprecise. Because we are imprecise by *convention*, we can make statements like

All automobiles are not alike.

and feel confident that the listener knows that we actually *mean*

Not all automobiles are alike.

We can also use spurious reasoning like

If it's raining then it's cloudy.
It is not raining.
Therefore there are no clouds.

and not expect to be challenged, because virtually everyone is careless when communicating informally. (Examples of this type will be considered in more detail later.)

Mathematics cannot tolerate this lack of rigor and precision. In order to achieve any depth beyond the most elementary level, we must adhere to strict rules of logic. The purpose of the present Appendix is to discuss the foundations of formal reasoning.

In this Appendix we shall often use numbers to illustrate logical concepts. The number systems we will encounter are

- The natural numbers $\mathbb{N} = \{1, 2, 3, \ldots\}$

- The integers $\mathbb{Z} = \{\ldots, -3, -2, -1, 0, 1, 2, 3, \ldots\}$

- The rational numbers $\mathbb{Q} = \{p/q : p \text{ is an integer}, q \text{ is an integer}, q \neq 0\}$

- The real numbers ℝ, consisting of all terminating and non-terminating decimal expansions. [This is a bit different from the way that we thought of the real numbers in Chapter 1, but it is logically equivalent.]

Chapter 1 reviewed the real and complex numbers. If you need to review the other number systems, then refer to Appendix I or [KRA1]. For now we assume that you have seen these number systems before. They are convenient for illustrating the logical principles we are discussing.

Section A2.1. "And" and "Or"

The statement

$$\text{``}\mathbf{A} \quad \text{and} \quad \mathbf{B}\text{''}$$

means that both **A** is true *and* **B** is true. For instance,

George is tall and George is intelligent.

means both that George is tall *and* George is intelligent. If we meet George and he turns out to be short and intelligent, then the statement is false. If he is tall and stupid then the statement is false. Finally, if George is *both* short and stupid then the statement is false. The statement is *true* precisely when both properties—intelligence and tallness—hold. We may summarize these assertions with a *truth table*. We let

A = George is tall.

and

B = George is intelligent.

The expression

$$\mathbf{A} \wedge \mathbf{B}$$

will denote the phrase "**A** and **B**." In particular, the symbol $\wedge$ is used to denote "and." The letters "T" and "F" denote "True" and "False," respectively. Then we have

A	B	A ∧ B
T	T	T
T	F	F
F	T	F
F	F	F

Notice that we have listed all possible truth values of **A** and **B** and the corresponding values of the *conjunction* $\mathbf{A} \wedge \mathbf{B}$.

In a restaurant the menu often contains phrases like

soup or salad

This means that we may select soup *or* select salad, but we may not select both. This use of "or" is called the *exclusive* "or"; it is not the meaning of "or" that we use in mathematics and logic. In mathematics we instead say that "**A or B**" is true provided that **A** is true or **B** is true or *both* are true. If we let **A** ∨ **B** denote "**A or B**" (the symbol ∨ denotes "or"), then the truth table is

A	B	A ∨ B
T	T	T
T	F	T
F	T	T
F	F	F

The only way that "**A** or **B**" can be false is if *both* **A** is false and **B** is false. For instance, the statement

Gary is handsome or Gary is rich.

means that Gary is either handsome or rich or both. In particular, he will not be both ugly and poor. Another way of saying this is that if he is poor he will compensate by being handsome; if he is ugly he will compensate by being rich. *But he could be both handsome and rich.*

Example A2.1

The statement

$$x > 5 \quad \text{and} \quad x < 7$$

is true for the number $x = 11/2$ because this value of x is both greater than 5 *and* less than 7. It is false for $x = 8$ because this x is greater than 5 but not less than 7. It is false for $x = 3$ because this x is less than 7 but not greater than 5. ∎

Example A2.2

The statement

x is even and x is a perfect square

is true for $x = 4$ because both assertions hold. It is false for $x = 2$ because this x, while even, is not a square. It is false for $x = 9$ because this x, while a square, is not even. It is false for $x = 5$ because this x is neither a square nor an even number. ∎

Example A2.3

The statement

$$x > 5 \quad \text{or} \quad x \leq 2$$

is true for $x = 1$ since this x is ≤ 2 (even though it is not > 5). It holds for $x = 6$ because this x is > 5 (even though it is not ≤ 2). The statement fails for $x = 3$ since this x is neither > 5 nor ≤ 2.   ∎

Example A2.4

The statement

$$x > 5 \quad \text{or} \quad x < 7$$

is true for every real x.   ∎

Example A2.5

The statement $(\mathbf{A} \vee \mathbf{B}) \wedge \mathbf{B}$ has the following truth table:

A	B	A ∨ B	(A ∨ B) ∧ B
T	T	T	T
T	F	T	F
F	T	T	T
F	F	F	F

∎

The words "and" and "or" are called *connectives*: their role in sentential logic is to enable us to build up (or connect together) pairs of statements. In the next section we will become acquainted with the other two basic connectives "not" and "if-then."

Section A2.2. "Not" and "If-Then"

The statement "not **A**," written $\sim \mathbf{A}$, is true whenever **A** is false. For example, the statement

Gene is not tall.

is true provided the statement "Gene is tall" is false. The truth table for $\sim \mathbf{A}$ is as follows

A	**$\sim$ A**
T	F
F	T

Although "not" is a simple idea, it can be a powerful tool when used in proofs by contradiction. To prove that a statement **A** is true using proof by contradiction, we instead assume $\sim \mathbf{A}$. We then show that this hypothesis leads to a contradiction. Thus $\sim \mathbf{A}$ must be false; according to the truth table, we see that the only possibility is that **A** is true.

Greater understanding is obtained by combining connectives:

Example A2.6

Here is the truth table for $\sim (\mathbf{A} \vee \mathbf{B})$:

A	**B**	**A $\vee$ B**	**$\sim$ (A $\vee$ B)**
T	T	T	F
T	F	T	F
F	T	T	F
F	F	F	T

∎

Example A2.7

Now we look at the truth table for $(\sim \mathbf{A}) \wedge (\sim \mathbf{B})$:

A	**B**	**$\sim$ A**	**$\sim$ B**	**$(\sim$ A$) \wedge (\sim$ B$)$**
T	T	F	F	F
T	F	F	T	F
F	T	T	F	F
F	F	T	T	T

∎

Notice that the statements $\sim (\mathbf{A} \vee \mathbf{B})$ and $(\sim \mathbf{A}) \wedge (\sim \mathbf{B})$ have the *same truth table*. We call such pairs of statements *logically equivalent*.

The logical equivalence of $\sim (\mathbf{A} \vee \mathbf{B})$ with $(\sim \mathbf{A}) \wedge (\sim \mathbf{B})$ makes good intuitive sense: the statement $\mathbf{A} \vee \mathbf{B}$ fails if and only if **A**   is false *and* **B** is false. Since in mathematics we cannot rely on our intuition to establish facts, it is important to have the truth table technique for establishing logical equivalence.

A statement of the form "If **A** then **B**" asserts that whenever **A** is true then **B** is also true. This assertion (or "promise") is tested when **A** is true, because it is then claimed that something else (namely, **B**) is true as well. *However*, when **A** is false, then the statement "If **A** then **B**" *claims nothing*. Using the symbols **A** ⇒ **B** to denote "If **A** then **B**," we obtain the following truth table:

A	**B**	**A** ⇒ **B**
T	T	T
T	F	F
F	T	T
F	F	T

Notice that we use here an important principle of Aristotelian logic: every sensible statement is either true or false. There is no "in between" status. Thus, when **A** is false, then the statement **A** ⇒ **B** is not tested. It therefore cannot be false. So it must be true. In fact the only way that **A** ⇒ **B** can be false is if **A** is true and **B** is false.

Example A2.8

The statement **A** ⇒ **B** is logically equivalent with ∼ (**A** ∧ ∼ **B**). For the truth table for the latter is

A	**B**	∼ **B**	**A** ∧ ∼ **B**	∼ (**A** ∧ ∼ **B**)
T	T	F	F	T
T	F	T	T	F
F	T	F	F	T
F	F	T	F	T

which is the same as the truth table for **A** ⇒ **B**.   ∎

There are in fact infinitely many pairs of logically equivalent statements. But just a few of these equivalences are really important in practice—most others are built up from these few basic ones.

Example A2.9

The statement

$$\text{If} \quad x \text{ is negative, then} \quad -5 \cdot x \text{ is positive.}$$

is true. For if $x < 0$, then $-5 \cdot x$ is indeed > 0; if $x \geq 0$, then the statement is unchallenged. ∎

Example A2.10

The statement

$$\text{If} \quad \{x > 0 \text{ and } x^2 < 0\}, \text{ then} \quad x \geq 10.$$

is true since the hypothesis "$x > 0$ and $x^2 < 0$" is never true.

∎

Example A2.11

The statement

$$\text{If} \quad x > 0, \text{ then } \{x^2 < 0 \text{ or } 2x < 0\}.$$

is false since the conclusion "$x^2 < 0$ or $2x < 0$" is false whenever the hypothesis $x > 0$ is true. ∎

Section A2.3. Contrapositive, Converse, and "Iff"

The statement

$$\text{If A then B.} \quad \text{or} \quad \text{A} \Rightarrow \text{B.}$$

is the same as saying

$$\text{A suffices for B.}$$

or as saying

$$\text{A only if B.}$$

All these forms are encountered in practice, and you should think about them long enough to realize that they all say the same thing.

On the other hand,

$$\text{If B then A.} \quad \text{or} \quad \text{B} \Rightarrow \text{A.}$$

is the same as saying

$$\text{A is necessary for B.}$$

or as saying

$$\textbf{A if B.}$$

We call the statement $\textbf{B} \Rightarrow \textbf{A}$ the *converse* of $\textbf{A} \Rightarrow \textbf{B}$.

Example A2.12

The converse of the statement

$$\textbf{If } x \textbf{ is a healthy horse, then } x \textbf{ has four legs.}$$

is the statement

$$\textbf{If } x \textbf{ has four legs, then } x \textbf{ is a healthy horse.}$$

Notice that these statements have very different meanings: the first statement is true while the second (its converse) is false. For example, my desk has four legs but it is not a healthy horse.

∎

The statement

$$\textbf{A if and only if B.}$$

is a brief way of saying

$$\textbf{If A then B.} \quad and \quad \textbf{If B then A.}$$

We abbreviate $\textbf{A}$ **if and only if** $\textbf{B}$ as $\textbf{A} \Leftrightarrow \textbf{B}$ or as $\textbf{A iff B.}$ Here is a truth table for $\textbf{A} \Leftrightarrow \textbf{B}$.

A	B	A $\Rightarrow$ B	B $\Rightarrow$ A	A $\Leftrightarrow$ B
T	T	T	T	T
T	F	F	T	F
F	T	T	F	F
F	F	T	T	T

Notice that we can say that $\textbf{A} \Leftrightarrow \textbf{B}$ is true only when both $\textbf{A} \Rightarrow \textbf{B}$ and $\textbf{B} \Rightarrow \textbf{A}$ are true. An examination of the truth table reveals that $\textbf{A} \Leftrightarrow \textbf{B}$ is true precisely when $\textbf{A}$ and $\textbf{B}$ are either both true or both false. Thus $\textbf{A} \Leftrightarrow \textbf{B}$ means precisely that $\textbf{A}$ and $\textbf{B}$ are logically equivalent. One is true when and *only when* the other is true.

Example A2.13

The statement

$$x > 0 \Leftrightarrow 2x > 0$$

is true. For if $x > 0$, then $2x > 0$; and if $2x > 0$, then $x > 0$. ∎

Example A2.14

The statement

$$x > 0 \Leftrightarrow x^2 > 0$$

is false. For $x > 0 \Rightarrow x^2 > 0$ is certainly true while $x^2 > 0 \Rightarrow x > 0$ is false ($(-3)^2 > 0$ but $-3 \not> 0$). ∎

Example A2.15

The statement

$$\{\sim (\mathbf{A} \vee \mathbf{B})\} \Leftrightarrow \{(\sim \mathbf{A}) \wedge (\sim \mathbf{B})\} \qquad \text{(A2.15.1)}$$

is true because the truth table for $\sim(\mathbf{A} \vee \mathbf{B})$ and that for $(\sim \mathbf{A}) \wedge (\sim \mathbf{B})$ are the same (we noted this fact in the last section). Thus they are logically equivalent: one statement is true precisely when the other is. Another way to see the truth of (A2.15.1) is to examine the truth table:

A	**B**	$\sim (\mathbf{A} \vee \mathbf{B})$	$(\sim \mathbf{A}) \wedge (\sim \mathbf{B})$	$\sim (\mathbf{A} \vee \mathbf{B}) \Leftrightarrow \{(\sim \mathbf{A}) \wedge (\sim \mathbf{B})\}$
T	T	F	F	T
T	F	F	F	T
F	T	F	F	T
F	F	T	T	T

∎

Given an implication

$$\mathbf{A} \Rightarrow \mathbf{B},$$

the *contrapositive* statement is defined to be the implication

$$\sim \mathbf{B} \Rightarrow \sim \mathbf{A}.$$

The contrapositive is logically equivalent to the original implication, as we see by examining their truth tables:

A	B	A $\Rightarrow$ B
T	T	T
T	F	F
F	T	T
F	F	T

and

A	B	$\sim$ A	$\sim$ B	($\sim$ B) $\Rightarrow$ ($\sim$ A)
T	T	F	F	T
T	F	F	T	F
F	T	T	F	T
F	F	T	T	T

Example A2.16

The statement

If it is raining, then it is cloudy.

has, as its contrapositive, the statement

If there are no clouds, then it is not raining.

A moment's thought convinces us that these two statements say the same thing: if there are no clouds, then it could not be raining; for the presence of rain implies the presence of clouds.

∎

Example A2.17

The statement

If X is a healthy horse, then X has four legs.

has, as its contrapositive, the statement

If X does not have four legs, then X is not a healthy horse.

A moment's thought reveals that these two statements say precisely the same thing. They are logically equivalent.

∎

The main point to keep in mind is that, given an implication $\mathbf{A} \Rightarrow \mathbf{B}$, its *converse* $\mathbf{B} \Rightarrow \mathbf{A}$ and its *contrapositive* $(\sim \mathbf{B}) \Rightarrow (\sim \mathbf{A})$ are two different statements. The converse is distinct from, and *logically independent from*, the original statement. The contrapositive is distinct from, but *logically equivalent to*, the original statement.

Section A2.4. Quantifiers

The mathematical statements that we will encounter in practice will use the *connectives* "and," "or," "not," "if–then," and "iff." They will also use *quantifiers*. The two basic quantifiers are "for all" and "there exists."

Example A2.18

Consider the statement

All automobiles have wheels.

This statement makes an assertion about *all* automobiles. It is true, just because every automobile does have wheels.

Compare this statement with the next one:

There exists a woman who is blonde.

This statement is of a different nature. It does not claim that all women have blonde hair—merely that there exists *at least one* woman who does. Since that is true, the statement is true. ∎

Example A2.19

Consider the statement

All positive real numbers are integers.

This sentence asserts that something is true for all positive real numbers. It is indeed true for *some* positive real numbers, such as 1 and 2 and 193. However, it is false for at least one positive number (such as π), so the entire statement is false.

Here is a more extreme example:

The square of any real number is positive.

This assertion is *almost* true—the only exception is the real number 0: we see that $0^2 = 0$ is not positive. But it only takes one exception to falsify a "for all" statement. So the assertion is false. ∎

Example A2.20

Look at the statement

There exists a real number which is greater than 4.

In fact there are lots of real numbers which are greater than 4; some examples are 7, 8π, and 97/3. Since there is *at least one* number satisfying the assertion, the assertion is true.

A somewhat different example is the sentence

There exists a real number which satisfies the equation
$$x^3 + x^2 + x + 1 = 0.$$

There is in fact only one real number which satisfies the equation, and that is $x = -1$. Yet that information is sufficient to make the statement true. ∎

We often use the symbol $\forall$ to denote "for all" and the symbol $\exists$ to denote "there exists." The assertion

$$\forall x, \ x + 1 < x$$

claims that, for every x, the number $x + 1$ is less than x. If we take our universe to be the standard real number system, this statement is false (for example, $5 + 1$ is not less than 5). The assertion

$$\exists x, \ x^2 = x$$

claims that there is a number whose square equals itself. If we take our universe to be the real numbers, then the assertion is satisfied by $x = 0$ and by $x = 1$. Therefore the assertion is true.

Quite often we will encounter $\forall$ and $\exists$ used together. The following examples are typical:

Example A2.21

The statement
$$\forall x \ \exists y, \ y > x$$

claims that for any number x there is a number y which is greater than it. In the realm of the real numbers this is true. In fact $y = x + 1$ will always do the trick.

The statement
$$\exists x \; \forall y, \; y > x$$
has quite a different meaning from the first one. It claims that there is an x which is less than *every* y. This is absurd. For instance, x is *not* less than $y = x - 1$. ∎

Example A2.22

The statement
$$\forall x \; \forall y, \; x^2 + y^2 \geq 0$$
is true in the realm of the real numbers: it claims that the sum of two squares is always greater than or equal to zero.

The statement
$$\exists x \; \exists y, \; x + 2y = 7$$
is true in the realm of the real numbers: it claims that there exist x and y such that $x + 2y = 7$. Certainly the numbers $x = 3, y = 2$ will do the job (although there are many other choices that work as well). ∎

We conclude by noting that $\forall$ and $\exists$ are closely related. The statements

$$\forall x, \; B(x) \qquad \text{and} \qquad {\sim}\,\exists x, \; {\sim} B(x)$$

are logically equivalent. The first asserts that the statement $B(x)$ is true for all values of x. The second asserts that there exists no value of x for which $B(x)$ fails, which is the same thing.

Likewise, the statements

$$\exists x, B(x) \qquad \text{and} \qquad {\sim}\,\forall x, \; {\sim} B(x)$$

are logically equivalent. The first asserts that there is some x for which $B(x)$ is true. The second claims that it is not the case that $B(x)$ fails for every x, which is the same thing.

Remark A2.23

Most of the statements that we encounter in mathematics are formulated using "for all" and "there exists." For example,

Through every point P not on a line ℓ there is a line parallel to ℓ.

Each continuous function on a closed, bounded interval has an absolute maximum.

Each of these statements uses (implicitly) both a "for all" and a "there exists."

A "for all" statement is like an *infinite conjunction*. The statement $\forall x, P(x)$ (when x is a natural number, let us say) says $P(1) \wedge P(2) \wedge P(3) \wedge \cdots$. A "there exists" statement is like an *infinite disjunction*. The statement $\exists x, Q(x)$ (when x is a natural number, let us say) says $Q(1) \vee Q(2) \vee Q(3) \vee \cdots$. Thus it is neither practical nor sensible to endeavor to verify statements such as these using truth tables. This is one of the chief reasons that we learn to produce mathematical proofs. One of the main themes of the present text is to gain new insights and to establish facts about the real number system using mathematical proofs.

Section A2.5. Set Theory and Venn Diagrams

The two most basic objects in all of mathematics are sets and functions. In this section we discuss the first of these two concepts.

A *set* is a collection of objects. For example, "the set of all blue shirts" and "the set of all lonely whales" are two examples of sets. In mathematics, we often write sets with the following "set-builder" notation:

$$\{x : x + 5 > 0\}.$$

This is read "the set of all x such that $x + 5$ is greater than 0." The universe from which x is chosen (for us this will usually be the real numbers) is understood from context, though sometimes we may be more explicit and write

$$\{x \in \mathbb{R} : x + 5 > 0\}.$$

Here $\in$ is a symbol that means "is an element of."

Notice that the role of x in the set-builder notation is as a *dummy variable*; the set we have just described could also be written as

$$\{s : s + 5 > 0\}$$

or

$$\{\alpha : \alpha + 5 > 0\}.$$

To repeat, the symbol $\in$ is used to express membership in a set; for example, the statement

$$4 \in \{x : x > 0\}$$

says that 4 is a member of (or *an element of*) the set of all numbers x which are greater than 0. In other words, 4 is a positive number.

If A and B are sets, then the statement

$$A \subset B$$

is read "A is a subset of B." It means that each element of A is also an element of B (but not vice versa!). In other words $x \in A \Rightarrow x \in B$.

Example A2.24

Let

$$A = \{x \in \mathbb{R} : \exists y \text{ such that } x = y^2\}$$

and

$$B = \{t \in \mathbb{R} : t + 3 > -5\}.$$

Then $A \subset B$. Why? The set A consists of those numbers that are squares— that is, A is just the nonnegative real numbers. The set B contains all numbers which are greater than -8. Since every nonnegative number (element of A) is also greater than -8 (element of B), it is correct to say that $A \subset B$.

However, it is not correct to say that $B \subset A$, because -2 is an element of B but is not an element of A. ∎

We write $A = B$ to indicate that both $A \subset B$ and $B \subset A$. In these circumstances we say that the two sets are equal: every element of A is an element of B and every element of B is an element of A.

We use a slash through the symbols $\in$ or $\subset$ to indicate negation:

$$-4 \notin \{x : x \geq -2\}$$

and

$$\{x : x = x^2\} \not\subset \{y : y > 1/2\}.$$

It is often useful to combine sets. The set $A \cup B$, called the *union* of A and B, is the set consisting of all objects which are either elements of A *or* elements of B (or both). The set $A \cap B$, called the *intersection* of A and B, is the set consisting of all objects which are elements of *both* A and B.

Example A2.25

Let

$$A = \{x : -4 < x \leq 3\} \quad, \quad B = \{x : -1 \leq x < 7\},$$
$$C = \{x : -9 \leq x \leq 12\}.$$

Then

$$A \cup B = \{x : -4 < x < 7\} \quad A \cap B = \{x : -1 \le x \le 3\},$$

$$B \cup C = \{x : -9 \le x \le 12\} \quad , \quad B \cap C = \{x : -1 \le x < 7\}.$$

Notice that $B \cup C = C$ and $B \cap C = B$ because $B \subset C$. ∎

Example A2.26

Let

$$A = \{\alpha \in \mathbb{Z} : \alpha \ge 9\}$$

$$B = \{\beta \in \mathbb{R} : -4 < \beta \le 24\},$$

$$C = \{\gamma \in \mathbb{R} : 13 < \gamma \le 30\}.$$

Then

$$(A \cap B) \cap C = \{x \in \mathbb{Z} : 9 \le x \le 24\} \cap C = \{t \in \mathbb{Z} : 13 < t \le 24\}.$$

Also

$$A \cap (B \cup C) = A \cap \{x \in \mathbb{R} : -4 < x \le 30\} = \{y \in \mathbb{Z} : 9 \le x \le 30\}.$$

Try your hand at calculating $A \cup (B \cup C)$. ∎

The symbol $\emptyset$ is used to denote the set with no elements. We call this set the *empty set*. For instance,

$$A = \{x \in \mathbb{R} : x^2 < 0\}$$

is a perfectly good set. However, there are no real numbers which satisfy the given condition. Thus A is empty, and we write $A = \emptyset$.

Example A2.27

Let

$$A = \{x : x > 8\} \quad \text{and} \quad B = \{x : x^2 < 4\}.$$

Then $A \cup B = \{x : x > 8 \text{ or } -2 < x < 2\}$ while $A \cap B = \emptyset$. ∎

We sometimes use a *Venn diagram* to aid our understanding of set-theoretic relationships. In a Venn diagram, a set is represented as a domain in the plane. The intersection $A \cap B$ of two sets A and B is the region common to the two domains—see Figure A2.1.

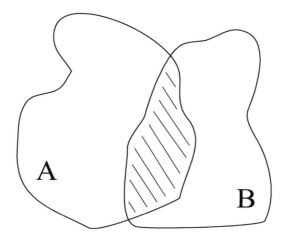

Figure A2.1: The intersection of two sets.

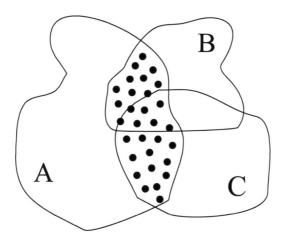

Figure A2.2: $A \cap (B \cup C) = (A \cap B) \cup (A \cap C)$.

Now let A, B, and C be three sets. The Venn diagram in Figure A2.2 makes it easy to see that $A \cap (B \cup C) = (A \cap B) \cup (A \cap C)$.

If A and B are sets then $A \setminus B$ denotes those elements which are *in A* but *not in B*. This operation is sometimes called *subtraction of sets* or *set-theoretic difference*.

Example A2.28

Let
$$A = \{x : 4 < x\}$$

and
$$B = \{x : 6 \leq x \leq 8\}.$$

Then
$$A \setminus B = \{x : 4 < x < 6\} \cup \{x : 8 < x\}$$

while
$$B \setminus A = \emptyset.$$

Notice that $A \setminus A = \emptyset$; this fact is true for any set. ∎

Example A2.29

Let
$$S = \{x : 5 \leq x\}$$

and
$$T = \{x : 4 < x < 6\}.$$

Then
$$S \setminus T = \{x : 6 \leq x\} \qquad \text{and} \qquad T \setminus S = \{x : 4 < x < 5\}.$$ ∎

The Venn diagram in Figure A2.3 illustrates the fact that
$$A \setminus (B \cup C) = (A \setminus B) \cap (A \setminus C)$$

A Venn diagram is not a proper substitute for a rigorous mathematical proof. However, it can go a long way toward guiding our intuition.

We conclude this section by mentioning a useful set-theoretic operation and an application. Suppose that we are studying subsets of a fixed set X. We sometimes call X the "universal set." If $S \subset X$ then we use the notation $^c S$ to denote the set $X \setminus S$ or $\{x \in X : x \notin S\}$. The set $^c S$ is called *the complement of S* (in the set X).

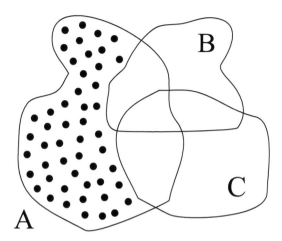

Figure A2.3: $A \setminus (B \cup C) = (A \setminus B) \cap (A \setminus C)$.

Example A2.30

When we study real analysis, most sets that we consider are subsets of the real line $\mathbb{R}$. If $S = \{x \in \mathbb{R} : 0 \leq x \leq 5\}$ then $^cS = \{x \in \mathbb{R} : x < 0\} \cup \{x \in \mathbb{R} : x > 5\}$. If T is the set of rational numbers then ^cT is the set of irrational numbers. ∎

If A, B are sets then it is straightforward to verify that $^c(A \cup B) = {}^cA \cap {}^cB$ and $^c(A \cap B) = {}^cA \cup {}^cB$. These are known as *de Morgan's laws*. Let us prove the first of these.

If $x \in {}^c(A \cup B)$ then x is not an element of $A \cup B$. Hence x is not an element of A and x is not an element of B. So $x \in {}^cA$ and $x \in {}^cB$. Therefore $x \in {}^cA \cap {}^cB$. That shows that $^c(A \cup B) \subset {}^cA \cap {}^cB$. For the reverse direction, assume that $x \in {}^cA \cap {}^cB$. Then $x \in {}^cA$ and $x \in {}^cB$. As a result, $x \notin A$ and $x \notin B$. So $x \notin A \cup B$. So $x \in {}^c(A \cup B)$. This shows that $^cA \cap {}^cB \subset {}^c(A \cup B)$. The two inclusions that we have proved establish that $^c(A \cup B) = {}^cA \cap {}^cB$.

Section A2.6. Relations and Functions

In more elementary mathematics courses we learn that a "relation" is a rule for associating elements of two sets; and a "function" is a rule that associates to each element of one set a unique element of another set. The trouble with these definitions is that they are imprecise. For example, suppose

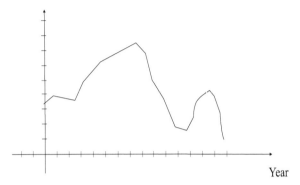

Figure A2.4: Value of the Yen against the Dollar.

we define the function $f(x)$ to be identically equal to 1 if there is life as we know it on Mars and to be identically equal to 0 if there is no life as we know it on Mars. Is this a good definition? It certainly is not a very practical one!

More important is the fact that using the word "rule" suggests that functions are given by formulas. Indeed, some functions are; but most are not. Look at any graph in the newspaper—of unemployment, or the value of the Japanese Yen (Figure A2.4), or the Gross National Product. The graphs represent values of these parameters as a function of time. And it is clear that the functions are not given by elementary formulas.

To summarize, we need a notion of function, and of relation, which is precise and flexible and which does not tie us to formulas. We begin with relations, and then specialize down to functions.

Definition A2.31

Let A and B be sets. A *relation* on A and B is a collection of ordered pairs (a, b) such that $a \in A$ and $b \in B$. (Notice that we did not say "*the* collection of all ordered pairs"—that is, a relation consists of some of the ordered pairs, but not necessarily all of them.)

Example A2.32

Let A be the real numbers and B the integers. The set

$$\mathcal{R} = \{(\pi, 2), (3.4, -2), (\sqrt{2}, 94), (\pi, 50), (2 + \sqrt{17}, -2)\}$$

is a relation on A and B. It associates certain elements of A to certain elements of B. Observe that repetitions are allowed: $\pi \in A$ is associated to both 2 and 50 in B; also $-2 \in B$ is associated to both 3.4 and $2 + \sqrt{17}$ in A. This relation is certainly not given by any formula or rule.

Now let

$$A = \{3, 17, 28, 42\} \quad \text{and} \quad B = \{10, 20, 30, 40\}.$$

Then

$$\mathcal{R} = \{(3, 10), (3, 20), (3, 30), (3, 40), (17, 20), (17, 30),$$
$$(17, 40), (28, 30), (28, 40)\}$$

is a relation on A and B. In fact $a \in A$ is related to $b \in B$ precisely when $a < b$. This second relation *is* given by a rule. ∎

Example A2.33

Let

$$A = B = \{\text{meter, pound, foot, ton, yard, ounce}\}.$$

Then

$$\mathcal{R} = \{(\text{foot,meter}), (\text{foot,yard}), (\text{meter,yard}), (\text{pound,ton}),$$
$$(\text{pound,ounce}), (\text{ton,ounce}), (\text{meter,foot}), (\text{yard,foot}),$$
$$(\text{yard,meter}), (\text{ton,pound}), (\text{ounce,pound}), (\text{ounce,ton})\}$$

is a relation on A and B. In fact two words are related by $\mathcal{R}$ if and only if they measure the same thing: foot, meter, and yard measure length while pound, ton, and ounce measure weight.

Notice that the pairs in $\mathcal{R}$, and in any relation, are *ordered* pairs: the pair (foot,yard) is different from the pair (yard,foot). ∎

Example A2.34

Let

$$A = \{25, 37, 428, 695\} \quad \text{and} \quad B = \{14, 7, 234, 999\}$$

Then

$$\mathcal{R} = \{(25, 234), (37, 7), (37, 234), (428, 14), (428, 234), (695, 999)\}$$

is a relation on A and B. In fact two elements are related by $\mathcal{R}$ if and only if they have at least one digit in common. ∎

Definition A2.35

Let $\mathcal{R}$ be a relation on a set A. We say that $\mathcal{R}$ is an *equivalence relation* if the following properties hold:

$\mathcal{R}$ **is reflexive:** If $x \in A$, then $(x, x) \in \mathcal{R}$;

$\mathcal{R}$ **is symmetric:** If $(x, y) \in \mathcal{R}$, then $(y, x) \in \mathcal{R}$;

$\mathcal{R}$ **is transitive:** If $(x, y) \in \mathcal{R}$ and $(y, z) \in \mathcal{R}$, then $(x, z) \in \mathcal{R}$.

Proposition A2.36

Let $\mathcal{R}$ be an equivalence relation on a set A. If $x \in A$, then define

$$E_x \equiv \{y \in A : (x, y) \in \mathcal{R}\}.$$

We call the sets E_x the equivalence classes induced by the relation $\mathcal{R}$. If now s and t are any two elements of A, then either $E_s \cap E_t = \emptyset$ or $E_s = E_t$.

In summary, the set A is the pairwise disjoint union of the equivalence classes induced by the equivalence relation $\mathcal{R}$.

Before we prove this proposition, let us discuss for a moment what it means. Clearly every element $a \in A$ is contained in some equivalence class, for a is contained in E_a itself. The proposition tells us that the set A is in fact the pairwise disjoint union of these equivalence classes. We say that the equivalence classes *partition* the set A.

Proof of the Proposition: Let $s, t \in A$ and suppose that $E_s \cap E_t \neq \emptyset$. It is our job to prove that $E_s = E_t$ (think for a moment about the truth table for "or" so that you understand that we are doing the right thing).

Since $E_s \cap E_t \neq \emptyset$, there is an element $x \in E_s \cap E_t$. Then $x \in E_s$. Therefore, by definition, $(s, x) \in \mathcal{R}$. Likewise, $x \in E_t$. Thus $(t, x) \in \mathcal{R}$. By symmetry, it follows that $(x, t) \in \mathcal{R}$. Now transitivity tells us that, since $(s, x) \in \mathcal{R}$ and $(x, t) \in \mathcal{R}$, then $(s, t) \in \mathcal{R}$.

If y is any element of E_t, then $(t, y) \in \mathcal{R}$. Transitivity now implies that since $(s, t) \in \mathcal{R}$ and $(t, y) \in \mathcal{R}$, then $(s, y) \in \mathcal{R}$. Thus $y \in E_s$. We have shown that every element of E_t is an element of E_s. Thus $E_t \subset E_s$.

Reversing the roles of s and t, we find that $E_s \subset E_t$. It follows that $E_s = E_t$. This is what we wished to prove. $\qquad\square$

Example A2.37

Let A be the set of all people in the United States. If $x, y \in A$, then let us say that $(x, y) \in \mathcal{R}$ if x and y have the same surname (i.e., last name). Then $\mathcal{R}$ is an equivalence relation:

(i) $\mathcal{R}$ is reflexive since any person x has the same surname as him/her self.

(ii) $\mathcal{R}$ is symmetric since if x has the same surname as y, then y has the same surname as x.

(iii) $\mathcal{R}$ is transitive since if x has the same surname as y and y has the same surname as z, then x has the same surname as z.

Thus $\mathcal{R}$ is an equivalence relation. The equivalence classes are all those people with surname Smith, all those people with surname Herkimer, and so forth. ∎

Example A2.38

Let S be the set of all residents of the United States. If $x, y \in S$, then let us say that x is related to y (that is, $x \sim y$) if x and y have at least one biological parent in common. It is easy to see that this relation is reflexive and symmetric. It is *not* transitive, as children of divorced parents know too well. What this tells us (mathematically) is that the proliferation of divorce in our society does *not* lead to well-defined families. ∎

Example A2.39

Let S be the set of integers and let us say that x is related to y if $x - y$ is divisible by 2. It is easily checked that this is an equivalence relation.

It is easy to see that any even integer is related to any other even integer; also any odd integer is related to any other odd integer. So the equivalence classes are two: $\mathcal{E}$ the even integers and $\mathcal{O}$ the odd integers. ∎

Example A2.40

Let S be the set of integers and say that $x \sim y$ if $x \leq y$. This relation is clearly reflexive. It is *not* symmetric, as $3 \leq 5$ but $5 \not\leq 3$. You may check that it is transitive. But the failure of symmetry tells us that this is not an equivalence relation. ∎

A function is a special type of relation, as we shall now learn.

Definition A2.41

Let A and B be sets. A *function* from A to B is a relation $\mathcal{R}$ on A and B such that for each $a \in A$ there is one and only one pair $(a, b) \in \mathcal{R}$. We call A the *domain* of the function and we call B the *range*.[2]

Example A2.42

Let
$$A = \{1, 2, 3, 4\} \quad \text{and} \quad B = \{\alpha, \beta, \gamma, \delta\}.$$
Then
$$\mathcal{R} = \{(1, \gamma), (2, \delta), (3, \gamma), (4, \alpha)\}$$
is a function from A to B. Notice that there is precisely one pair in $\mathcal{R}$ for each element of A. However, repetition of elements of B is allowed. Observe also that there is no apparent "pattern" or "rule" that determines $\mathcal{R}$. Finally observe that not all the elements of B are used.

With the same sets A and B consider the relations
$$\mathcal{S} = \{(1, \alpha), (2, \beta), (3, \gamma)\}$$
and
$$\mathcal{T} = \{(1, \alpha), (2, \beta), (3, \gamma), (4, \delta), (2, \gamma)\}.$$
Then $\mathcal{S}$ is not a function because it violates the rule that there be a pair for *each* element of A. Also $\mathcal{T}$ is not a function because it violates the rule that there be *just one* pair for each element of A.   ∎

The relations and function described in the last example were so simple that you may be wondering what happened to the kinds of functions that we usually look at in mathematics. Now we consider some of those.

Example A2.43

Let $A = \mathbb{R}$ and $B = \mathbb{R}$, where $\mathbb{R}$ denotes the real numbers. The relation
$$\mathcal{R} = \{(x, \sin x) : x \in A\}$$
is a function from A to B. For each $a \in A = \mathbb{R}$ there is one and only one ordered pair with first element a.

---

[2]Some textbooks use the word "codomain" instead of range. We shall use only the word "range."

Now let $S = \mathbb{R}$ and $T = \{x \in \mathbb{R} : -2 \le x \le 2\}$. Then

$$\mathcal{U} = \{(x, \sin x) : x \in S\}$$

is also a function from S to T. Technically speaking, it is a different function from $\mathcal{R}$ because it has a different range. However, this distinction often has no practical importance and we shall not mention the difference. It is frequently convenient to write functions like $\mathcal{R}$ or $\mathcal{U}$ as

$$\mathcal{R}(x) = \sin x$$

and

$$\mathcal{U}(x) = \sin x \,.\qquad\blacksquare$$

The last example suggests that we distinguish between the set B where a function takes its values and the set of values that the function *actually assumes*.

Definition A2.44

Let A and B be sets and let f be a function from A to B. Define the *image* of f to be

$$\text{Image}\, f = \{b \in B : \exists a \in A \text{ such that } f(a) = b\}\,.$$

The set Image f is a subset of the range B. In general the image *will not* equal the range.

Example A2.45

Both the functions $\mathcal{R}$ and $\mathcal{U}$ from the last example have the set $\{x \in \mathbb{R} : -1 \le x \le 1\}$ as image. In neither instance does the image equal the range. $\blacksquare$

If a function f has domain A and range B and if S is a subset of A then we define

$$f(S) = \{b \in B : b = f(s) \text{ for some } s \in S\}\,.$$

The set $f(A)$ equals the image of f. $\blacksquare$

Example A2.46

Let $A = \mathbb{R}$ and $B = \{0, 1\}$. Consider the function

$$f = \{(x, y) : y = 0 \text{ if } x \text{ is rational and}$$

$$y = 1 \text{ if } x \text{ is irrational}\}\,.$$

The function f is called the *Dirichlet function* (P. G. Lejeune-Dirichlet, 1805-1859). It is given by a rule, but not by a formula.

Notice that $f(\mathbb{Q}) = \{0\}$ and $f(\mathbb{R}) = \{0,1\}$. ∎

Definition A2.47

Let A and B be sets and f a function from A to B.

We say that f is *one-to-one* if whenever $(a_1, b) \in f$ and $(a_2, b) \in f$ then $a_1 = a_2$.

We say that f is *onto* if whenever $b \in B$ then there exists an $a \in A$ such that $(a, b) \in f$.

Example A2.48

Let $A = \mathbb{R}$ and $B = \mathbb{R}$. Consider the functions

$$f(x) = 2x + 5 \quad , \quad g(x) = \arctan x$$

$$h(x) = \sin x \quad , \quad j(x) = 2x^3 + 9x^2 + 12x + 4\,.$$

Then f is both one-to-one and onto, g is one-to-one but not onto, j is onto but not one-to-one, and h is neither.

Refer to Figure A2.5 to convince yourself of these assertions. ∎

When a function f is both one-to-one and onto then it is called a *bijection* of its domain to its range. Sometimes we call such a function a *set-theoretic isomorphism* or *bijection*. In the last example, the function f is a bijection of $\mathbb{R}$ to $\mathbb{R}$.

If f and g are functions, and if the image of g is contained in the domain of f, then we define the *composition* $f \circ g$ to be

$$\{(a, c) : \exists b \text{ such that } g(a) = b \text{ and } f(b) = c\}\,.$$

This may be written more simply as

$$f \circ g(a) = f(g(a)) = f(b) = c\,.$$

Let f have domain A and range B. Assume for simplicity that the image of f is all of B. If there exists a function g with domain B and range A such that

$$f \circ g(b) = b \quad \forall b \in B$$

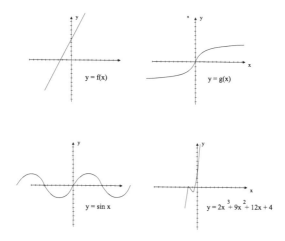

Figure A2.5: One-to-one and onto functions.

and

$$g \circ f(a) = a \quad \forall a \in A,$$

then g is called the *inverse* of f. We frequently write f^{-1} for the inverse function.

Clearly, if the function f is to have an inverse, then f must be one-to-one. For if $f(a) = f(a') = b$ then it cannot be that both $g(b) = a$ and $g(b) = a'$. Also f must be onto. For if some $b \in B$ is not in the image of f then it cannot hold that $f \circ g(b) = b$. It turns out that these two conditions are also sufficient for the function f to have an inverse: If f has domain A and range B and if f is both one-to-one and onto, then f has an inverse.

Example A2.49

Define a function f, with domain $\mathbb{R}$ and range $\{x \in \mathbb{R} : x \geq 0\}$, by the formula $f(x) = x^2$. Then f is onto but is not one-to-one (because $f(-1) = f(1)$), hence it cannot have an inverse. This is another way of saying that a positive real number has two square roots—not one.

However, the function g, with domain $\{x \in \mathbb{R} : x \geq 0\}$ and range $\{x \in \mathbb{R} : x \geq 0\}$, given by the formula $g(x) = x^2$, *does* have an inverse. In fact the inverse function is $h(x) = +\sqrt{x}$.

The function $k(x) = x^3$, with domain $\mathbb{R}$ and range $\mathbb{R}$, is both one-to-one and onto. It therefore has an inverse: the function $m(x) = x^{1/3}$ satisfies the condition $k \circ m(x) = x$, and also $m \circ k(x) = x$ for all x. ∎

Section A2.7. Countable and Uncountable Sets

One of the most profound ideas of modern mathematics is Georg Cantor's theory of the infinite (Georg Cantor, 1845-1918). Cantor's insight was that infinite sets can be compared by size, just as finite sets can. For instance, we think of the number 2 as *less* than the number 3; so a set with two elements is "smaller" than a set with three elements. We would like to have a similar notion of comparison for infinite sets. In this section we will present Cantor's ideas; we will also give precise definitions of the terms "finite" and "infinite."

Definition A2.50

Let A and B be sets. We say that A and B have the *same cardinality* if there is a function f from A to B which is both one-to-one and onto (that is, f is a bijection or set-theoretic isomorphism from A to B). We write $\mathrm{card}(A) = \mathrm{card}(B)$. Some books write $|A| = |B|$.

Example A2.51

Let $A = \{1, 2, 3, 4, 5\}, B = \{\alpha, \beta, \gamma, \delta, \epsilon\}, C = \{a, b, c, d, e, f\}$. Then A and B have the same cardinality because the function

$$f = \{(1, \alpha), (2, \beta), (3, \gamma), (4, \delta), (5, \epsilon)\}$$

is a bijection of A to B. This function is not the *only* bijection of A to B (can you find another?), but we are only required to produce one.

On the other hand, A and C do not have the same cardinality; neither do B and C. ∎

Notice that if $\mathrm{card}(A) = \mathrm{card}(B)$ via a function f_1 and $\mathrm{card}(B) = \mathrm{card}(C)$ via a function f_2 then $\mathrm{card}(A) = \mathrm{card}(C)$ via the function $f_2 \circ f_1$.

Example A2.52

Let A and B be sets. If there is a one-to-one function from A to B but no bijection between A and B then we will write

$$\mathrm{card}(A) < \mathrm{card}(B).$$

This notation is read "A has smaller cardinality than B."

We use the notation

$$\text{card}(A) \leq \text{card}(B)$$

to mean that either $\text{card}(A) < \text{card}(B)$ or $\text{card}(A) = \text{card}(B)$. ∎

Example A2.53

An extremely simple example of this last concept is given by $A = \{1, 2, 3\}$ and $B = \{a, b, c, d, e\}$. Then the function

$$
\begin{array}{rcl}
f : A & \rightarrow & B \\
1 & \mapsto & a \\
2 & \mapsto & b \\
3 & \mapsto & c
\end{array}
$$

is a one-to-one function from A to B. But there is no one-to-one function from B to A. We write

$$\text{card}(A) < \text{card}(B).$$

We shall see more profound applications, involving infinite sets, in our later discussions. ∎

Notice that $\text{card}(A) \leq \text{card}(B)$ and $\text{card}(B) \leq \text{card}(C)$ imply that $\text{card}(A) \leq \text{card}(C)$. Moreover, if $A \subset B$, then the inclusion map $i(a) = a$ is a one-to-one function of A into B; therefore $\text{card}(A) \leq \text{card}(B)$.

The next theorem gives a useful method for comparing the cardinality of two sets.

Theorem A2.54 (Schroeder-Bernstein)

Let A, B, be sets. If there is a one-to-one function $f : A \rightarrow B$ and a one-to-one function $g : B \rightarrow A$, then A and B have the same cardinality.

Proof: It is convenient to assume that A and B are disjoint; we may do so by replacing A by $\{(a, 0) : a \in A\}$ and B by $\{(b, 1) : b \in B\}$. Let D be the image of f and C be the image of g. Let us define a *chain* to be a sequence of elements of either A or B—that is, a function $\phi : \mathbb{N} \rightarrow (A \cup B)$—such that

- $\phi(1) \in B \setminus D$;

- If for some j we have $\phi(j) \in B$, then $\phi(j + 1) = g(\phi(j))$;

- If for some j we have $\phi(j) \in A$, then $\phi(j + 1) = f(\phi(j))$.

We see that a chain is a sequence of elements of $A \cup B$ such that the first element is in $B \setminus D$, the second in A, the third in B, and so on. Obviously each element of $B \setminus D$ occurs as the first element of at least one chain.

Define $S = \{a \in A : a$ is some term of some chain$\}$. It is helpful to note that

$$S = \{x : x \text{ can be written in the form}$$
$$g(f(g(\cdots g(y)\ldots))) \text{ for some } y \in B \setminus D\}. \qquad (A2.54.1)$$

We set

$$k(x) = \begin{cases} f(x) & \text{if } x \in A \setminus S \\ g^{-1}(x) & \text{if } x \in S \end{cases}$$

Note that the second half of this definition makes sense because $S \subseteq C$. Then $k : A \to B$. We shall show that in fact k is a bijection.

First notice that f and g^{-1} are one-to-one. This is not quite enough to show that k is one-to-one, but we now reason as follows: If $f(x_1) = g^{-1}(x_2)$ for some $x_1 \in A \setminus S$ and some $x_2 \in S$, then $x_2 = g(f(x_1))$. But, by (A2.54.1), the fact that $x_2 \in S$ now implies that $x_1 \in S$. That is a contradiction. Hence k is one-to-one.

It remains to show that k is onto. Fix $b \in B$. We seek an $x \in A$ such that $k(x) = b$.

Case A: If $g(b) \in S$, then $k(g(b)) \equiv g^{-1}(g(b)) = b$ hence the x that we seek is $g(b)$.

Case B: If $g(b) \notin S$, then we claim that there is an $x \in A$ such that $f(x) = b$. Assume this claim for the moment.

Now the x that we found in the last paragraph must lie in $A \setminus S$. For if not then x would be in some chain. Then $f(x)$ and $g(f(x)) = g(b)$ would also lie in that chain. Hence $g(b) \in S$, and that is a contradiction. But $x \in A \setminus S$ tells us that $k(x) = f(x) = b$. That completes the proof that k is onto. Hence k is a bijection.

To prove the claim in Case B, notice that if there is no x with $f(x) = b$, then $b \in B \setminus D$. Thus some chain would begin at b. So $g(b)$ would be a term of that chain. Hence $g(b) \in S$ and that is a contradiction.

The proof of the Schroeder-Bernstein theorem is complete. $\qquad\qquad \square$

Remark A2.55

Let us reiterate some of the earlier ideas in light of the Schroeder-Bernstein theorem. If A and B are sets and if there is a one-to-one function $f : A \to B$, then we know that $\text{card}(A) \leq \text{card}(B)$. If there is no one-to-one function

$g : B \rightarrow A$, then we may write $\mathrm{card}(A) < \mathrm{card}(B)$. But if instead there *is* a one-to-one function $g : B \rightarrow A$, then $\mathrm{card}(B) \leq \mathrm{card}(A)$ and the Schroeder-Bernstein theorem guarantees therefore that $\mathrm{card}(A) = \mathrm{card}(B)$.

Now it is time to look at some specific examples.

Example A2.56

Let E be the set of all even integers and O the set of all odd integers. Then
$$\mathrm{card}(E) = \mathrm{card}(O).$$
Indeed, the function
$$f(j) = j + 1$$
is a bijection from E to O. ∎

Example A2.57

Let E be the set of even integers. Then
$$\mathrm{card}(E) = \mathrm{card}(\mathbb{Z}).$$
The function
$$g(j) = j/2$$
is a bijection from E to $\mathbb{Z}$. ∎

This last example is a bit surprising, for it shows that a set $\mathbb{Z}$ can be put in one-to-one correspondence with a proper subset E of itself. In other words, we are saying that the integers $\mathbb{Z}$ "have the same number of elements" as a proper subset of $\mathbb{Z}$. Such a phenomenon cannot occur with finite sets.

Example A2.58

We have
$$\mathrm{card}(\mathbb{Z}) = \mathrm{card}(\mathbb{N}).$$
We define the function f from $\mathbb{Z}$ to $\mathbb{N}$ as follows:

- $f(j) = -(2j + 1)$ if j is negative

- $f(j) = 2j + 2$ if j is positive or zero

The values that f takes on the negative numbers are $1, 3, 5, \ldots$, on the positive numbers are $4, 6, 8, \ldots$, and $f(0) = 2$. Thus f is one-to-one and onto. ∎

Definition A2.59

If a set A has the same cardinality as $\mathbb{N}$ then we say that A is *countable*.

By putting together the preceding examples, we see that the set of even integers, the set of odd integers, and the set of all integers are examples of countable sets.   ∎

Example A2.60

The set of all ordered pairs of positive integers

$$S = \{(j, k) : j, k \in \mathbb{N}\}$$

is countable.

To see this we will use the Schroeder-Bernstein theorem. The function

$$f(j) = (j, 1)$$

is a one-to-one function from $\mathbb{N}$ to S. Also the function $g(j, k) = 2^j \cdot 3^k$ is a one-to-one function from S to $\mathbb{N}$. By the Schroeder-Bernstein theorem, S and $\mathbb{N}$ have the same cardinality; hence S is countable.   ∎

Remark A2.61

You may check for yourself that the function $F(j, k) = 2^{j-1} \cdot (2k - 1)$ is an explicit bijection from S to $\mathbb{N}$.

Since there is a bijection of the set of *all* integers with the set $\mathbb{N}$, it follows from the last example that the set of all pairs of integers (positive *and* negative) is countable.

Notice that the word "countable" is a good descriptive word: if S is a countable set then we can think of S as having a first element (the one corresponding to $1 \in \mathbb{N}$), a second element (the one corresponding to $2 \in \mathbb{N}$), and so forth. Thus we write $S = \{s(1), s(2), \ldots\} = \{s_1, s_2, \ldots\}$.

Definition A2.62

A nonempty set S is called *finite* if there is a bijection of S with a set of the form $\{1, 2, \ldots, n\}$ for some positive integer n. If no such bijection exists, then the set is called *infinite*.

An important property of the natural numbers $\mathbb{N}$ is that any subset $S \subset \mathbb{N}$ has a least element. This is known as the Well Ordering Principle, and is studied in a course on logic. In the present text we take the properties of the natural numbers as given. We use some of these properties in the next proposition.

Proposition A2.63

If S is a countable set and R is a subset of S then either R is empty or R is finite or R is countable.

Proof: Assume that R is not empty.

Write $S = \{s_1, s_2, \ldots\}$. Let j_1 be the least positive integer such that $s_{j_1} \in R$. Let j_2 be the least integer following j_1 such that $s_{j_2} \in R$. Continue in this fashion. If the process terminates at the nth step, then R is finite and has n elements.

If the process does not terminate, then we obtain an enumeration of the elements of R:

$$1 \longleftrightarrow s_{j_1}$$
$$2 \longleftrightarrow s_{j_2}$$
$$\cdots$$

etc.

All elements of R are enumerated in this fashion since $j_\ell \geq \ell$. Therefore R is countable. $\qquad\square$

A set is called *denumerable* if it is either empty, finite, or countable. Notice that the word "denumerable" is not the same as "countable." In fact "countable" is just one instance of denumerable.

The set $\mathbb{Q}$ of all rational numbers consists of all expressions

$$\frac{a}{b},$$

where a and b are integers and $b \neq 0$. Thus $\mathbb{Q}$ can be identified with the set of all ordered pairs (a, b) of integers with $b \neq 0$. After discarding duplicates, such as $\frac{2}{4} = \frac{1}{2}$, and using Examples A2.58 and A2.60 and Proposition A2.63, we find that the set $\mathbb{Q}$ is countable.

Theorem A2.64

Let S_1, S_2 be countable sets. Set $S = S_1 \cup S_2$. Then S is countable.

Proof: Let us write
$$S_1 = \{s_1^1, s_2^1, \ldots\}$$
$$S_2 = \{s_1^2, s_2^2, \ldots\}.$$

If $S_1 \cap S_2 = \emptyset$ then the function

$$s_j^k \mapsto (j, k)$$

is a bijection of S with a subset of $\{(j,k) : j, k \in \mathbb{N}\}$. We proved earlier (Example A2.60) that the set of ordered pairs of elements of $\mathbb{N}$ is countable. By Proposition A2.63, S is countable as well.

If there exist elements which are common to S_1, S_2 then discard any duplicates. The same argument (use the preceding proposition) shows that S is countable. $\square$

Theorem A2.65

If S and T are each countable sets then so is

$$S \times T \equiv \{(s, t) : s \in S, t \in T\}.$$

Proof: Since S is countable there is a bijection f from S to $\mathbb{N}$. Likewise there is a bijection g from T to $\mathbb{N}$. Therefore the function

$$(f \times g)(s, t) = (f(s), g(t))$$

is a bijection of $S \times T$ with $\mathbb{N} \times \mathbb{N}$, the set of order pairs of positive integers. But we saw in Example A2.60 that the latter is a countable set. Hence so is $S \times T$. $\square$

Remark A2.66

We take note of the concept of *set-theoretic product*: If A and B are sets then

$$A \times B \equiv \{(a, b) : a \in A, b \in B\}.$$

More generally, if $A_1, A_2, \ldots, A_k$ are sets then

$$A_1 \times A_2 \times \cdots \times A_k \equiv \{(a_1, a_2, \ldots, a_k) : a_j \in A_j \text{ for all } j = 1, \ldots, k\}.$$

Corollary A2.67

If $S_1, S_2, \ldots, S_k$ are each countable sets then so is the set

$$S_1 \times S_2 \times \cdots \times S_k = \{(s_1, \ldots, s_k) : s_1 \in S_1, \ldots, s_k \in S_k\}$$

consisting of all ordered $k-$tuples $(s_1, s_2, \ldots, s_k)$ with $s_j \in S_j$.

Proof: We may think of $S_1 \times S_2 \times S_3$ as $(S_1 \times S_2) \times S_3$. Since $S_1 \times S_2$ is countable (by Theorem A2.60) and S_3 is countable, then so is $(S_1 \times S_2) \times S_3 = S_1 \times S_2 \times S_3$ countable. Continuing in this fashion, we can see that any finite product of countable sets is also a countable set. □

We are accustomed to the union $A \cup B$ of two sets or, more generally, the union $A_1 \cup A_2 \cup \cdots \cup A_k$ of finitely many sets. But sometimes we wish to consider the union of infinitely many sets. Let $S_1, S_2, \ldots$ be countably many sets. We say that x is an element of

$$\bigcup_{j=1}^{\infty} S_j$$

if x is an element of at least one of the S_j.

Corollary A2.68

The countable union of countable sets is countable.

Proof: Let $A_1, A_2, \ldots$ each be countable sets. If the elements of A_j are enumerated as $\{a_k^j\}$ and if the sets A_j are pairwise disjoint then the correspondence

$$a_k^j \longleftrightarrow (j, k)$$

is one-to-one between the union of the sets A_j and the countable set $\mathbb{N} \times \mathbb{N}$. This proves the result when the sets A_j have no common element. If some of the A_j have elements in common then we discard duplicates in the union and use Proposition A2.63. □

Proposition A2.69

The collection $\mathcal{P}$ of all polynomials with integer coefficients is countable.

Proof: Let $\mathcal{P}_k$ be the set of polynomials of degree k with integer coefficients. A polynomial p of degree k has the form

$$p(x) = p_0 + p_1 x + p_2 x^2 + \cdots + p_k x^k .$$

The identification

$$p(x) \longleftrightarrow (p_0, p_1, \ldots, p_k)$$

identifies the elements of $\mathcal{P}_k$ with the $(k+1)$-tuples of integers. By Corollary A2.67, it follows that $\mathcal{P}_k$ is countable. But then Corollary A2.68 implies that

$$\mathcal{P} = \bigcup_{j=0}^{\infty} \mathcal{P}_j$$

is countable.   □

Georg Cantor's remarkable discovery is that *not all infinite sets are countable.* We next give an example of this phenomenon.

In what follows, a *sequence* on a set S is a function from $\mathbb{N}$ to S. We usually write such a sequence as $s(1), s(2), s(3), \ldots$ or as $s_1, s_2, s_3, \ldots$.

Example A2.70

There exists an infinite set which is not countable (we call such a set *uncountable*). Our example will be the set S of all sequences on the set $\{0, 1\}$. In other words, S is the set of all infinite sequences of 0s and 1s. To see that S is uncountable, assume the contrary. Then there is a first sequence

$$S^1 = \{s_j^1\}_{j=1}^{\infty} ,$$

a second sequence

$$S^2 = \{s_j^2\}_{j=1}^{\infty} ,$$

and so forth. This will be a complete enumeration of all the members of S. But now consider the sequence $\mathcal{T} = \{t_j\}_{j=1}^{\infty}$, which we construct as follows:

- If $s_1^1 = 0$ then make $t_1 = 1$; if $s_1^1 = 1$ then set $t_1 = 0$;

- If $s_2^2 = 0$ then make $t_2 = 1$; if $s_2^2 = 1$ then set $t_2 = 0$;

- If $s_3^3 = 0$ then make $t_3 = 1$; if $s_3^3 = 1$ then set $t_3 = 0$;

. . .

- If $s_j^j = 0$ then make $t_j = 1$; if $s_j^j = 1$ then make $t_j = 0$;

etc.

Now the sequence $\mathcal{T}$ differs from the first sequence $\mathcal{S}^1$ in the first element: $t_1 \neq s_1^1$.

The sequence $\mathcal{T}$ differs from the second sequence $\mathcal{S}^2$ in the second element: $t_2 \neq s_2^2$.

And so on: the sequence $\mathcal{T}$ differs from the jth sequence $\mathcal{S}^j$ in the jth element: $t_j \neq s_j^j$. So the sequence $\mathcal{T}$ is not in the set S. But $\mathcal{T}$ is *supposed* to be in the set S because it is a sequence of 0s and 1s and all of these have been hypothesized to be enumerated.

This contradicts our assumption, so S must be uncountable.

∎

Example A2.71

Consider the set of all decimal representations of numbers—both terminating and non-terminating. Here a terminating decimal is one of the form

$$27.43926$$

while a non-terminating decimal is one of the form

$$3.14159265\ldots.$$

In the case of the non-terminating decimal, no repetition is implied; the decimal simply continues without cease.

Now the set of all those decimals containing only the digits 0 and 1 can be identified in a natural way with the set of sequences containing only 0 and 1 (just put commas between the digits). And we just saw that the set of such sequences is uncountable.

Since the set of all decimal numbers is an even bigger set, it must be uncountable also.

As you may know, the set of all decimals identifies with the set of all real numbers. We find then that the set $\mathbb{R}$ of all real numbers is uncountable. (Contrast this with the situation for the rationals.) In Chapter 1 we learn more about how the real number system is constructed using just elementary set theory. ∎

It is an important result of set theory (due to Cantor) that, given any set S, the set of all subsets of S (called the *power set* of S) has strictly greater cardinality than the set S itself. As a simple example, let $S = \{a, b, c\}$. Then the set of all subsets of S is

$$\left\{ \, \emptyset, \{a\}, \{b\}, \{c\}, \{a, b\}, \{a, c\}, \{b, c\}, \{a, b, c\} \, \right\}.$$

The set of all subsets has eight elements while the original set has just three.

Even more significant is the fact that, if S is an infinite set, then the set of all its subsets has greater cardinality than S itself. This is a famous theorem of Cantor. Thus there are infinite sets of arbitrarily large cardinality.

In some of the examples in this Appendix we constructed a bijection between a given set (such as $\mathbb{Z}$) and a proper subset of that set (such as E, the even integers). It follows from the definitions that this is possible only when the sets involved are infinite.

Table of Notation

Notation	Section	Definition
$\mathbb{Q}$	1.1	the rational numbers
$\sup X$	1.1	supremum of X
$\operatorname{lub} X$	1.1	least upper bound of X
$\inf X$	1.1	infimum of X
$\operatorname{glb} X$	1.1	greatest lower bound of X
$\mathbb{R}$	1.1	the real numbers
$\lvert x \rvert$	1.1	absolute value
$\lvert x+y \rvert \leq \lvert x \rvert + \lvert y \rvert$	1.1	triangle inequality
$\mathcal{P}$	1.1	a cut
$\mathbb{C}$	1.2	the complex numbers
z	1.2	a complex number
i	1.2	the square root of -1
$\overline{z}$	1.2	complex conjugate
$\lvert z \rvert$	1.2	modulus of z
$e^{i\theta}$	1.2	complex exponential
$\{a_j\}$	2.1	a sequence
a_j	2.1	a sequence
a_{j_k}	2.2	a subsequence
$\liminf a_j$	2.3	limit infimum of a_j
$\limsup a_j$	2.3	limit supremum of a_j
a^j	2.4	a power sequence
e	2.4	Euler's number e
$\sum_{j=1}^{\infty} a_j$	3.1	a series
S_N	3.1	a partial sum
$\sum_{j=1}^{N} a_j$	3.1	a partial sum
$\sum_{j=1}^{\infty} (-1)^j b_j$	3.3	an alternating series

Notation	Section	Definition
$j!$	3.4	j factorial
$\sum_{n-0}^{\infty} \sum_{j=0}^{n} a_j \cdot b_{n-j}$	3.5	the Cauchy product of series
(a, b)	4.1	open interval
$[a, b]$	4.1	closed interval
$[a, b)$	4.1	half-open interval
$(a, b]$	4.1	half-open interval
U	4.1	an open set
F	4.1	a closed set
∂S	4.1	boundary of S
$^c S$	4.1	complement of S
$\overline{S}$	4.2	closure of S
$\overset{\circ}{S}$	4.2	interior of S
$\{\mathcal{O}_\alpha\}$	4.3	an open cover
S_j	4.4	step in constructing the Cantor set
C	4.4	the Cantor set
$\lim_{E \ni x \to c} f(x)$	5.1	limit of f at c
ℓ	5.1	a limit
$f + g$	5.1	sum of functions
$f - g$	5.1	difference of functions
$f \cdot g$	5.1	product of functions
f/g	5.1	quotient of functions
$f \circ g$	5.2	composition of functions
f^{-1}	5.2	inverse function
$f^{-1}(W)$	5.2	inverse image of a set
$f(L)$	5.3	image of the set L
m	5.3	minimum for a function f
M	5.3	maximum for a function f
$\lim_{x \to c^-} f(x)$	5.4	left limit of f at c
$\lim_{x \to c^+} f(x)$	5.4	right limit of f at c
$f'(x)$	6.1	derivative of f at x
df/dx	6.1	derivative of f
$\dot{f}$	6.1	derivative of f
$\text{Lip}_\alpha(I)$	6.3	space of Lipschitz functions
$C^{k,\alpha}(I)$	6.3	space of smooth functions of order k, α
$\mathcal{P}$	7.1	a partition
I_j	7.1	interval from the partition
Δ_j	7.1	length of I_j
$m(\mathcal{P})$	7.1	mesh of the partition
$\mathcal{R}(f, \mathcal{P})$	7.1	Riemann sum

Notation	Section	Definition
$\int_a^b f(x)\,dx$	7.1	Riemann integral
$\int_b^a f(x)\,dx$	7.2	integral with reverse orientation
f_j	8.1	sequence of functions
$\{f_j\}$	8.1	sequence of functions
$\lim_{x \to s} f(x)$	8.2	limit of f as x approaches s
$\sum_{j=1}^\infty f_j(x)$	8.3	series of functions
$S_N(x)$	8.3	partial sum of a series of functions
$p(x)$	8.4	a polynomial
$\sum_{j=0}^\infty a_j(x-c)^j$	9.1	a power series
R_N	9.1	tail of the power series
ρ	9.2	radius of convergence
$f(x) = \sum_{j=0}^k f^{(j)}(a)\frac{(x-a)^j}{j!}$ $+ R_{k,a}(x)$	9.2	Taylor expansion
$\exp(x)$	9.3	the exponential function
$\sin x$	9.3	the sine function
$\cos x$	9.3	the cosine function
$\mathrm{Sin}\,x$	9.3	sine with restricted domain
$\mathrm{Cos}\,x$	9.3	cosine with restricted domain
$\ln x$		the natural logarithm function
$\binom{n}{k}$	A1.1	binomial coefficient
$\mathbb{Z}$	A1.2	the integers
$[(a,b)]$	A1.2	an integer
$\mathbb{Q}$	A1.3	the rational numbers
$[(c,d)]$	A1.3	a rational number
$\wedge$	A2.1	the connective "and"
$\vee$	A2.1	the connective "or"
$\sim$	A2.2	the connective "not"
$\Rightarrow$	A2.2	the connective "if-then"
$\{\}$	A2.5	a set
$\Leftrightarrow$	A2.3	the connective "if and only if"

Notation	Section	Definition
iff	A2.3	the connective "if and only if"
$\forall$	A2.4	the quantifier "for all"
$\exists$	A2.4	the quantifier "there exists"
$\in$	A2.5	is an element of
$\subset$	A2.5	subset of
$\notin$	A2.5	is not an element of
$\cap$	A2.5	intersection
$\cup$	A2.5	union
$\emptyset$	A2.5	the empty set
$\setminus$	A2.5	set-theoretic difference
^{c}S	A2.5	complement of the set S
(a,b)	A2.6	a relation
$f(x)$	A2.6	a function
$f \circ g$	A2.6	composition of functions
$\mathrm{card}(A)$	A2.7	the cardinality of A
$|A|$	A2.7	the cardinality of A
E	A2.7	the even integers
O	A2.7	the odd integers
$\times$	A2.7	set-theoretic product

GLOSSARY

Abel's convergence test A test for convergence of series that is based on summation by parts.

absolutely convergent series A series for which the absolute values of the terms form a convergent series.

absolute maximum A number M is the absolute maximum for a function f if $f(x) \leq f(M)$ for every x.

absolute minimum A number m is the absolute minimum for a function f if $f(x) \geq f(m)$ for every x.

absolute value Given a real number x, its absolute value is the distance of x to 0.

accumulation point A point x is an accumulation point of a set S if every neighborhood of x contains infinitely many distinct elements of S.

alternating series A series of real terms which alternate in sign.

alternating series test If an alternating series has terms tending to zero then it converges.

"and" The connective which is used for conjunction.

Archimedean Property If a and b are positive real numbers then there is a positive integer n so that $na > b$.

bijection A one-to-one, onto function.

binomial expansion   The expansion, under multiplication, of the expression $(a + b)^n$.

Bolzano-Weierstrass Theorem   Every bounded sequence of real numbers has a convergent subsequence.

boundary point   The point b is in the boundary of S if each neighborhood of b contains both points of S and points of the complement of S.

boundary of a set   The set of boundary points for the set.

bounded above   A subset $S \subset \mathbb{R}$ is bounded above if there is a real number b such that $s \leq b$ for all $s \in S$.

bounded below   A subset $S \subset \mathbb{R}$ is bounded below if there is a real number c such that $s \geq c$ for all $s \in S$.

bounded sequence   A sequence a_j with the property that there is a number M so that $|a_j| \leq M$ for every j.

bounded set   A set S with the property that there is a number M with $|s| \leq M$ for every $s \in S$.

Cantor set   A compact set which is uncountable, has zero length, is perfect, is totally disconnected, and has many other unusual properties.

cardinality   Two sets have the same cardinality when there is a one-to-one correspondence between them.

Cauchy Condensation Test   A series of decreasing, nonnegative terms converges if and only if its dyadically condense series converges.

Cauchy criterion   A sequence a_j is said to be Cauchy if, for each $\epsilon > 0$, there is an $N > 0$ so that, if $j, k > N$, then $|a_j - a_k| < \epsilon$.

Cauchy criterion for a series   A series satisfies the Cauchy criterion if and only if the sequence of partial sums satisfies the Cauchy criterion for a sequence.

Cauchy product A means for taking the product of two series.

Cauchy's Mean Value Theorem A generalization of the Mean Value Theorem that allows the comparison of two functions.

Chain Rule A rule for differentiating the composition of functions.

change of variable A method for transforming an integral by subjecting the domain of integration to a one-to-one function.

closed set The complement of an open set.

closure of a set The set together with its boundary points.

common refinement of two partitions The union of the two distinct partitions.

compact set A set E is compact if every sequence in E contains a subsequence that converges to an element of E.

Comparison Test for Convergence A series converges if it is majorized in absolute value by a convergent series.

Comparison Test for Divergence A series diverges if it majorizes a divergent series.

complement of a set The set of points not in the set.

complex conjugate Given a complex number $z = x + iy$, the conjugate is the number $\bar{z} = x - iy$.

complex numbers The set $\mathbb{C}$ of ordered pairs of real numbers equipped with certain operations of addition and multiplication.

composition The composition of two functions is the succession of one function by the other.

conditionally convergent series A series which converges, but not absolutely.

connected set   A set which cannot be separated by two disjoint open sets.

connectives   The words which are used to connect logical statements. These are "and," "or," "not," "if-then," and "if and only if."

continuity at a point   The function f is continuous at c if the limit of f at c equals the value of f at c. Equivalently, given $\epsilon > 0$, there is a $\delta > 0$ so that $|x - c| < \delta$ implies $|f(x) - f(c)| < \epsilon$.

continuous function   A function which is continuous at each point c in the domain.

continuously differentiable function   A function which has a derivative at every point, and so that the derivative function is continuous.

convergence of a series   A series converges if and only if its sequence of partial sums converges.

convergence of a sequence (of numbers)   A sequence a_j with the property that there is a limiting element ℓ so that, for any $\epsilon > 0$, there is a positive integer N so that, if $j > n$, then $|a_j - \ell| < \epsilon$.

converse   For a statement "**A implies B**", the converse statement is "**B implies A**".

contrapositive   For a statement "**A implies B**", the contrapositive statement is "**$\sim$ B implies $\sim$ A**".

cosine function   The function $\cos x = \sum_{j=0}^{\infty} (-1)^j x^{2j}/(2j)!$.

countable set   Any set that has precisely the same cardinality as the natural numbers.

decreasing sequence   The sequence of real numbers a_j is decreasing if $a_1 \geq a_2 \geq a_3 \geq \cdots$.

Dedekind cut   A rational halfline that is bounded above in $\mathbb{Q}$. Used to construct the real numbers.

de Morgan's Laws The identities

$$^c(A \cup B) = {}^cA \cap {}^cB$$

and

$$^c(A \cap B) = {}^cA \cup {}^cB.$$

Density Property If $c < d$ are real numbers then there is a rational number q with $c < q < d$.

denumerable set A set that is either empty, finite, or countable.

derivative The limit $\lim_{t \to x}(f(x) - f(t))/(t - x)$ for a function f on an open interval.

derived power series The series obtained by differentiating a power series term by term.

difference quotient The quotient $(f(t) - f(x))/(t - x)$ for a function f on an open interval.

differentiable A function that possesses the derivative at a point.

Dirichlet function A function, taking only the values 0 and 1, which is highly discontinuous.

disconnected set A set which can be separated by two disjoint open sets.

discontinuity of the first kind A point at which a function f is discontinuous because the left and right limits at the point disagree.

discontinuity of the second kind A point at which a function f is discontinuous because either the left limit or the right limit at the point does not exist.

diverge to infinity A sequence with elements that become arbitrarily large.

domain of a function See *function*.

domain of integration   The interval over which the integration is performed.

dummy variable   A variable whose role in an argument or expression is formal. A dummy variable can be replaced by any other variable with no logical consequences.

element of   A member of a given set.

empty set   The set with no elements.

equivalence classes   The pairwise disjoint sets into which an equivalence relation partitions a set.

equivalence relation   A relation on a set S that is reflexive, symmetric, and transitive. It partitions the set into equivalence classes.

Euler's formula   The identity $e^{iy} = \cos y + i \sin y$.

Euler's number   This is the number $e = 2.71828\ldots$ which is known to be irrational, indeed transcendental.

exponential function   The function $\exp(z) = \sum_{j=0}^{\infty} z^j / j!$.

field   A system of numbers equipped with operations of addition and multiplication and satisfying eleven natural axioms.

finite set   A set that can be put in one-to-one correspondence with a set of the form $\{1, 2, \ldots, n\}$ for some positive integer n.

"for all"   The quantifier $\forall$ for making a statement about all objects of a certain kind.

function   A *function* from a set A to a set B is a relation f on A and B such that for each $a \in A$ there is one and only one pair $(a, b) \in f$. We call A the *domain* and B the *range* of the function.

Fundamental Theorem of Calculus   A result relating the values of a function to the integral of its derivative: $f(x) - f(a) = \int_a^x f'(t)\, dt$.

geometric series   This is a series of powers.

greatest lower bound   The real number c is the greatest lower bound for the set $S \subset \mathbb{R}$ if b is a lower bound and if there is no lower bound that is greater than c.

i The square root of -1 in the complex number system.

if   An alternative phrase for converse implication.

"if and only if"   The connective which is used for logical equivalence.

"if-then"   The connective which is used for implication.

image of a function   See *function*. The image of the function f is Image $f = \{b \in B : \exists a \in A \text{ such that } f(a) = b\}$.

image of a set   If f is a function then the image of E under f is the set $\{f(e) : e \in E\}$.

imaginary part   Given a complex number $z = x + iy$, its imaginary part is y.

increasing sequence   The sequence of real numbers a_j is increasing if $a_1 \leq a_2 \leq a_3 \leq \cdots$.

infimum   See *greatest lower bound*.

infinite set   A set is infinite if it is not finite.

integers   The natural numbers, the negatives of the natural numbers, and zero.

integration by parts   A device for integrating a product.

interior of a set   The collection of interior points of the set.

interior point   A point of the set S which has a neighborhood lying in S.

intermediate value theorem   The result that says that a continuous function does not skip values.

intersection of sets   The set of elements common to two or more given sets.

interval   A subset of the reals that contains all its intermediate points.

interval of convergence of a power series   An interval of the form $(c - \rho, c + \rho)$ on which the power series converges (uniformly on compact subsets of the interval).

irrational number   A real number which is not rational.

isolated point of a set   A point of the set with a neighborhood containing no other point of the set.

k times continuously differentiable   A function that has k derivatives, each of which is continuous.

Lambert W function   A transcendental function W with the property that any of the standard transcendental functions (sine, cosine, exponential, logarithm) can be expressed in terms of W.

least upper bound   The real number b is the least upper bound for the set $S \subset \mathbb{R}$ if b is an upper bound and if there is no other upper bound that is less than b.

Least Upper Bound Property   The important defining property of the real numbers.

left limit   A limit of a function at a point c that is calculated with values of the function that are to the left of c.

higher derivatives   The derivative of a derivative.

l'Hôpital's Rule   A rule for calculating the limit of the quotient of two functions in terms of the quotient of the derivatives.

limit   The value ℓ that a function approaches at a point of or an accumulation point c of the domain. Equivalently, given $\epsilon > 0$, there is a $\delta > 0$ so that

$|f(x) - \ell| < \epsilon$ whenever $|x - c| < \delta$.

limit infimum   The least limit of any subsequence of a given sequence.

limit supremum   The greatest limit of any subsequence of a given sequence.

Lipschitz function   A function that satisfies a condition of the form $|f(s) - f(t)| \leq C|s - t|$ or $|f(s) - f(t)| \leq |s - t|^\alpha$ for $0 < \alpha \leq 1$.

local extrema   Either a local maximum or a local minimum.

local maximum   The point x is a local maximum for the function f if $f(x) \geq f(t)$ for all t in a neighborhood of x.

local minimum   The point x is a local minimum for the function f if $f(x) \leq f(t)$ for all t in a neighborhood of x.

logically equivalent   Two statements are logically equivalent if they have the same truth table.

logically independent   Two statements are logically independent if neither one implies the other.

lower bound   A real number c is an lower bound for a subset $S \subset \mathbb{R}$ if $s \geq c$ for all $s \in S$.

Mean Value Theorem   If f is a continuous function on $[a, b]$, differentiable on the interior, then the slope of the segment connecting $(a, f(a))$ and $(b, f(b))$ equals the derivative of f at some interior point.

mesh of a partition   The maximum length of any interval in the partition.

modulus   The modulus of a complex number $z = x + iy$ is $|z| = \sqrt{x^2 + y^2}$.

monotone sequence   A sequence that is either increasing or decreasing.

monotonic function   A function that is either monotonically increasing or monotonically decreasing.

monotonically decreasing function   A function whose graph goes downhill when moving from left to right: $f(s) \geq f(t)$ when $s < t$.

monotonically increasing function   A function whose graph goes uphill when moving from left to right: $f(s) \leq f(t)$ when $s < t$.

natural logarithm function   The inverse function to the exponential function.

natural numbers   The counting numbers $1, 2, 3, \ldots$.

necessary for   An alternative phrase for converse implication.

neighborhood of a point   An open set containing the point.

Neumann series   A series of the form $1/(1 - \alpha) = \sum_{j=0}^{\infty} \alpha^j$ for $|\alpha| < 1$.

Newton quotient   The quotient $(f(t) - f(x))/(t - x)$ for a function f on an open interval.

non-terminating decimal expansion   A decimal expansion for a real number that has infinitely many nonzero digits.

"not"   The connective which is used for negation.

one-to-one function   A function that takes different values at different points of the domain.

only if   An alternative phrase for implication.

onto function   A function whose image equals its range.

open ball   The set of points at distance less than some $r > 0$ from a fixed point c.

open set   A set which contains a neighborhood of each of its points.

"or"   The connective which is used for disjunction.

ordered field   A field equipped with an order relation that is compatible with the field structure.

partial sum of functions   The sum of the first N terms of a series of functions.

partial sum (of scalars)   The sum of the first N terms of a series of scalars.

partition of the interval $[a, b]$ A finite, ordered set of points $\mathcal{P} = \{x_0, x_1, x_2, \ldots, x_{k-1}, x_k\}$ such that

$$a = x_0 \leq x_1 \leq x_2 \leq \cdots \leq x_{k-1} \leq x_k = b.$$

Peano axioms   An axiom system for the natural numbers.

perfect set   A set which is closed and in which every point is an accumulation point.

Pinching Principle   A criterion for convergence of a sequence that involves bounding it below by a convergent sequence and bounding it above by another convergent sequence with the same limit.

pointwise convergence of a sequence of functions   A sequence f_j of functions converges pointwise if $f_j(x)$ convergence for each x in the common domain.

polar form of a complex number   The polar form of a complex number z is $re^{i\theta}$, where r is the modulus of z and θ is the angle that the vector from 0 to z subtends with the positive x-axis.

power series expanded about the point c   A series of the form

$$\sum_{j=0}^{\infty} a_j(x - c)^j.$$

vspace*.1in
power set   The collection of all subsets of a given set.

Principle of Induction   A proof technique for establishing a statement $Q(n)$ about the natural numbers.

quantifier   A logical device for making a quantitative statement. Our standard quantifiers are "for all" and "there exists."

radius of convergence of a power series   Half the length ρ of the interval of convergence.

range of a functon   See *function*.

rational numbers   Numbers which may be represented as quotients of integers.

Ratio Test for Convergence   A series converges if the limit of the sequence of quotients of summands is less than 1.

Ratio Test for Divergence   A series diverges if the limit of the sequence of quotients of summands is greater than 1.

real-analytic function   A function with a convergent power series expansion about each point of its domain.

real numbers   An ordered field $\mathbb{R}$ containing the rationals $\mathbb{Q}$ so that every nonempty subset with an upper bound has a least upper bound.

real part   Given a complex number $z = x + iy$, its real part is x.

rearrangement of a series   A new series obtained by permuting the summands of the original series.

relation   A relation on sets A and B is a subset of $A \times B$.

remainder term for the Taylor expansion   The term $R_{k,a}(x)$ in the Taylor expansion.

Riemann integrable   A function for which the Riemann integral exists.

Riemann integral   The limit of the Riemann sums.

Riemann sum The approximate integral based on a partition.

right limit A limit of a function at a point c that is calculated with values of the function to the right of c.

Rolle's Theorem The special case of the Mean Value Theorem when $f(a) = f(b) = 0$.

Root Test for Convergence A series converges if the limit of the nth roots of the nth terms is less than one.

Root Test for Divergence A series is divergent if the limit of the nth roots of the nth terms is greater than one.

scalar An element of either $\mathbb{R}$ or $\mathbb{C}$.

Schroeder-Bernstein Theorem The result that says that if there is a one-to-one function from the set A to the set B and a one-to-one function from the set B to the set A then A and B have the same cardinality.

sequence of functions A function from $\mathbb{N}$ into the set of functions on some space.

sequence (of scalars) A function from $\mathbb{N}$ into $\mathbb{R}$ or $\mathbb{C}$ or a metric space. We often denote the sequence by a_j.

series of functions An infinite sum of functions.

series (of scalars) An infinite sum of scalars.

set A collection of objects.

setbuilder notation The notation $\{x : c(x)\}$ for specifying a set.

set-theoretic difference The set-theoretic difference $A \setminus B$ consists of those elements that lie in A but not in B.

set-theoretic isomorphism A one-to-one, onto function.

set-theoretic product   If A and B are sets then their set-theoretic product is the set of ordered pairs (a, b) with $a \in A$ and $b \in B$.

sine function   The function $\sin x = \sum_{j=0}^{\infty} (-1)^j x^{2j+1}/(2j+1)!$.

smaller cardinality   The set A has smaller cardinality than the set B if there is a one-to-one mapping of A to B but none from B to A.

strictly monotonically decreasing function   A function whose graph goes strictly downhill when moving from left to right: $f(s) > f(t)$ when $s < t$.

strictly monotonically increasing function   A function whose graph goes strictly uphill when moving from left to right: $f(s) < f(t)$ when $s < t$.

subfield   Given a field k, a subfield m is a subset of k which is also a field with the induced field structure.

subsequence   A sequence that is a subset of a given sequence with the elements occurring in the same order.

subset of   A subcollection of the members of a given set.

successor   The natural number which follows a given natural number.

suffices for   An alternative phrase for implication.

summation by parts   A discrete analogue of integration by parts.

supremum   See *least upper bound*.

Taylor's expansion   The expansion $f(x) = \sum_{j=0}^{k} f^{(j)}(a) \frac{(x-a)^j}{j!} + R_{k,a}(x)$ for a given function f.

terminating decimal   A decimal expansion for a real number that has only finitely many nonzero digits.

"there exists"   The quantifier $\exists$ for making a statement about some objects of a certain kind.

totally disconnected set   A set in which any two points can be separated by two disjoint open sets.

transcendental number   A real number which is not algebraic.

triangle inequality   The inequality

$$|a + b| \leq |a| + |b|$$

for real numbers a and b.

truth table   An array which shows the possible truth values of a statement.

uncountable set   A set that does not have the same cardinality as the natural numbers.

uniform convergence of a sequence of functions   The sequence f_j of functions converges uniformly to a function f if, given $\epsilon > 0$, there is an $N > 0$ so that, if $j > N$, then $|f_j(x) - f(x)| < \epsilon$ for all x.

uniform convergence of a series of functions   A series of functions such that the sequence of partial sums converges uniformly.

uniformly Cauchy sequence of functions   A sequence of functions f_j with the property that, for $\epsilon > 0$, there is an $N > 0$ so that, if $j, k > N$, then $|f_j(x) - f_k(x)| < \epsilon$ for all x in the common domain.

uniformly continuous   A function f is uniformly continuous if, for each $\epsilon > 0$, there is a $\delta > 0$ so that $|f(s) - f(t)| < \epsilon$ whenever $|s - t| < \delta$.

union of sets   The collection of objects that lie in any one of a given collection of sets.

universal set   The set of which all other sets are a subset.

upper bound   A real number b is an upper bound for a subset $S \subset \mathbb{R}$ if $s \leq b$ for all $s \in S$.

Venn diagram   A pictorial device for showing relationships among sets.

Weierstrass Approximation Theorem   The result that any continuous function on $[0, 1]$ can be uniformly approximated by polynomials.

Weierstrass M-Test   A simple scalar test that guarantees the uniform convergence of a series of functions.

Weierstrass Nowhere Differentiable Function   A function that is continuous on $[0, 1]$ that is not differentiable at any point of $[0, 1]$.

well defined   An operation on equivalence classes is well defined if the result is independent of the representatives chosen from the equivalence classes.

Zero Test   If a series converges then its summands tend to zero.

Bibliography

[**BOA1**] R. C. Boas, *A Primer of Real Functions*, Carus Mathematical Monograph No. 13, John Wiley & Sons, Inc., New York, 1960.

[**BUC**] R. C. Buck, *Advanced Calculus*, 2d ed., McGraw-Hill Book Company, New York, 1965.

[**FED**] H. Federer, *Geometric Measure Theory*, Springer-Verlag, New York, 1969.

[**HOF**] K. Hoffman, *Analysis in Euclidean Space*, Prentice Hall, Inc., Englewood Cliffs, NJ, 1962.

[**KRA1**] S. G. Krantz, *The Elements of Advanced Mathematics*, 2nd ed., CRC Press, Boca Raton, FL, 2002.

[**KRA4**] S. G. Krantz, *Handbook of Logic and Proof Techniques for Computer Scientists*, Birkhäuser, Boston, 2002.

[**KRA5**] S. G. Krantz, *Real Analysis and Foundations*, 3rd ed., Taylor & Francis, Boca Raton, FL, 2013.

[**KRA6**] S. G. Krantz, *Function Theory of Several Complex Variables*, 2nd ed., American Mathematical Society, Providence, RI, 2001.

[**KRP**] S. G. Krantz and H. R. Parks, *A Primer of Real Analytic Functions*, 2nd ed., Birkhäuser Publishing, Boston, 2002.

[**LOS**] L. Loomis and S. Sternberg, *Advanced Calculus*, Addison-Wesley, Reading, MA, 1968.

[**NIV**] I. Niven, *Irrational Numbers*, Carus Mathematical Monograph No. 11, John Wiley & Sons, Inc., New York, 1956.

[**ROY**] H. Royden, *Real Analysis*, Macmillan, New York, 1963.

[**RUD1**] W. Rudin, *Principles of Mathematical Analysis*, 3rd ed., McGraw-Hill Book Company, New York, 1976.

[**RUD2**] W. Rudin, *Real and Complex Analysis*, McGraw-Hill Book Company, New York, 1966.

[**STRO**] K. Stromberg, *An Introduction to Classical Real Analysis*, Wadsworth Publishing, Inc., Belmont, CA, 1981.

Index

Abel's Convergence Test, 59
absolute convergence, 62
absolute convergence of series, 62
absolute maximum, 115
absolute minimum, 115
absolute value, 6
absolutely convergent series, 62
accumulation point, 81
addition, 6, 228
addition of integers, 222
addition of rational numbers, 227
addition of series, 69
additive identity, 228
additive identity for rational numbers, 228
additive inverse, 228
algebra
    fundamental theorem of, 16
$\alpha^{1/j}$, 36
alternating series
    rate of convergence, 60
Alternating Series Test, 60, 64
"and", 236, 245
approximation of a continuous function by a polynomial, 186
Archimedean Property of the Real Numbers, 5
Aristotelian logic, 240
associativity of addition, 225, 228
associativity of multiplication, 225, 228
axioms for a field, 6, 228
axioms for the natural numbers, 217

Bernoulli, J., 142

bijection, 260, 272
binomial coefficient, 220
binomial theorem, 66
binomimal formula, 220
Bolzano-Weierstrass theorem, 28, 86
boundary of a set, 82
boundary point, 82
bounded set, 86
boundedness of a continuous function on a compact set, 114

Cantor set, 90, 94
    in terms of series, 92
    is uncountable, 92
Cantor ternary set, 90, 93
Cantor, G., 262, 270
cardinality of a set, 262
Cauchy Condensation Test, 49
Cauchy criterion for series, 44
Cauchy product of series, 70
Cauchy sequence, 22
    converges, 23
    is bounded, 23
Cauchy's Mean Value Theorem, 140
Cauchy, A. L., 140
Chain Rule, 132
change of variable, 162
characterization of connected subsets of $\mathbb{R}$, 96
characterization of open sets of reals, 78
$C^{k,\alpha}$ function, 146
closed intervals, 79
closed set

characterization of in terms of sequences, 80
closed sets, 79
    intersection of, 80
closure of a set, 83
closure of addition, 228
closure of multiplication, 228
codomain of a function, 258
coefficients of a power series, 202
combining sets, 249
common refinement of partitions, 154
commutativity of addition, 225, 228
commutativity of multiplication, 225, 228
commuting limits, 175
compact sets, 87, 88
    compared to finite sets, 88
comparison of the Root and Ratio

Tests, 54
Comparison Test, 48
complement, 252
complex conjugate, 15
complex number
    modulus of, 15
    polar form, 16
complex number system
    properties of, 12
complex numbers, 11
    addition of, 11
    additive identity for, 12
    multiplication of, 11
    not an ordered field, 17
    product of, 209
    properties of, 14
    standard notation for, 14
composition of functions, 260
conditional convergence, 62
conditionally convergence of series, 62
connected set, 95
connected sets are intervals, 96
connectives, 238
continuity, 108

elementary properties of, 110
continuity and sequences, 110
continuity under composition, 111
continuous function, 108
    discontinuities of, 123
continuous functions
    characterization using sequences, 110
continuous functions are integrable, 155
continuous image of a compact set, 113
continuously differentiable, 145
contradiction
    proof by, 239
contrapositive, 241, 243, 244
convergence
    absolute, 62
    conditional, 62
convergence of a series of functions, 180
convergence of Taylor series
    counterexample to, 204
convergence tests
    cancellation, 58
convergence tests for series
    advanced, 57
    elementary, 48
convergence tests for series of positive terms, 48
converse, 241, 242, 244
cosine function, 209
    Taylor series for, 209
countable set, 262, 266
countable set of discontinuities, 123
countable union of countable sets, 268
cuts, 7
    addition of, 8
    multiplication of, 8

Darboux's Theorem, 136
de Morgan's laws, 253
decimal representation, 5
decreasing function, 122

decreasing sequence, 24
Dedekind cuts, 7, 13
Dedekind, Julius, 6
Density Property of the Real Numbers, 5
denumerable set, 267
derivation of the Weierstrass Nowhere Differentiable Function, 148
derivative, 127
    notation for, 128
derivative of the inverse function, 144
derived power series, 201
difference of integers, 223
differentiable, 127
differentiable implies continuous, 128
Dini's theorem, 183
Dirichlet function, 103, 157, 260
Dirichlet, P., 260
disconnected set, 95
discontinuities of a monotonic function, 123
discontinuities of derivative are of second kind, 136
discontinuity of the first kind, 121
discontinuity of the second kind, 121
distributive law, 228
divergence of a series of functions, 180
domain of a function, 258

element of a set, 248
elementary properties of the derivative, 129
elementary properties of the integral, 159
elements arbitrarily close to lim sup, 34
elements arbitrarily close to liminf, 34
empty set, 250
equivalence class, 221, 256
equivalence relation, 226, 256
Euler's formula, 209
Euler's number e, 37, 65

example of a non-real-analytic function, 204
exponential function, 206, 213, 214
    elementary properties of, 206, 208, 214
    properties of, 207
    Taylor series for, 206

"false", 236
Fermat's theorem, 135
field, 6, 228
    axioms for, 229
finite set, 262, 267
"for all", 245
"for all", 245–247
function, 253, 257
function as a set of ordered pairs, 254
functions, 248
Fundamental Theorem of Algebra, 16
Fundamental Theorem of Calculus, 164, 165

Gauss's lemma, 234
Gauss, K., 218
geometric series, 51, 198
greater cardinality, 262
Gronwall's inequality, 141

Hadamard formula for the radius of convergence, 200, 201
harmonic series, 51
Heine-Borel theorem, 88
higher derivatives, 145
Hölder continuity, 145
homeomorphism, 119

"if", 242
"if and only if", 242
"iff", 241, 245
"if–then", 238, 241, 245
image
    of a function, 259
    of a set, 259

image does not equal range, 259
image of a function, 113, 259
imprecision of everyday language, 235
improper integrals, 166
increasing function, 122
increasing sequence, 24
induction, 217
infimum, 2, 3
infinite set, 262, 267
integers, 221, 225, 235
    as equivalence classes, 222
    construction of, 222
integrable functions are bounded, 157
integral and area, 151
Integral Test, 56
interior point, 84
Intermediate Value Theorem, 117
intersection of closed sets, 80
intersection of nested compact sets, 89
intersection of sets, 249
interval
    closed, 75
    half-closed, 75
    half-open, 75
    open, 75
interval notation, 75
interval of convergence, 194
intervals are connected, 96
inverse of a function, 261
irrationality of $\sqrt{2}$, 233
isolated point, 84, 85

$j^{1/j}$, 36

k^{th} derivative, 145
k times continuously differentiable, 145
k times differentiable, 145

law of the excluded middle, 240
least upper bound, 2
least upper bound property, 9
Least Upper Bound Property of the Real Numbers, 3

Lebesgue integral, 151
left limit, 120
length of a set, 91
l'Hôpital's Rule, 134, 142, 144
library
    of basic sequences, 35
    of basic series, 65
lim inf, 31
limit inferior, 31
    characterizing property, 33
limit infimum, 31
limit of a function at a point, 101
limit of a sequence
    properties of, 21
limit of a sequence of differentiable functions, 177
limit of continuous functions is not continuous, 170
limit of Riemann sums, 153
limit of the greatest subsequence, 32
limit of the least subsequence, 32
limit superior, 31
    characterizing property, 33
limit supremum, 31
limits, 1, 101
    characterization using sequences, 107
    elementary properties of, 104
    uniqueness of, 104
limits do commute when the convergence is uniform, 175
limits do not commute, 175
limits of functions using sequences, 107
lim sup, 31
Lipschitz condition, 112, 184
Lipschitz continuity, 145
local maximum, 135
local minimum, 135
logarithm function, 213
    multiplicative property, 215
logically equivalent, 239, 243
logically equivalent statements, 240
lower bound

greatest, 3

maxima, 135
Mean Value Theorem, 134, 138
membership in a set, 248
mesh of a partition, 151
minima, 135
modulus of a complex number, 15
monotone, 122
monotone sequence, 24
monotonic, 122
monotonically decreasing, 123
monotonically increasing, 123
monotonicity, 122
multiplication, 6, 228
    of integers, 224
    of series, 70
multiplicative
    identity, 228
    inverse, 228

natural logarithm function, 213, 215
natural numbers, 217, 235
"necessary for", 242
negation, 249
neighborhood, 76
Newton quotient, 127
no holes, 2
nonisolated point, 86, 87
"not", 238, 245
nowhere differentiable function, 131
n^{th} root, existence of, 5
n^{th} roots of real numbers, 5
number π, 211
number systems, 217, 236

one-to-one, 260
one-to-one function, 260
"only if", 241
onto, 260
onto function, 260
open intervals, 78
open sets, 76

intersection of, 77
union of, 77
operations on series, 69
"or", 236, 237, 245
order relation, 232
ordered field, 3, 9, 232
ordering, 231
oscillatory discontinuity, 122

partial sum, 42
partition, 151
Peano axioms, 217
Peano, G., 217
perfect set, 98
perfect sets are uncountable, 99
π
    definition of, 211
Pinching Principle, 25
polar form of a complex number, 16
polynomial of degree k has k roots, 16
polynomials
    are continuous, 108
    are differentiable, 130
    with integer coefficients, 270
power sequences, 35
    rational and real exponents, 35
power series, 193
    differentiation of, 201, 202
    integration of, 202
power series representation
    uniqueness of, 202
power set, 272
Principle of Induction, 218
product of
    countable sets, 269
    integers, 224
    integrable functions, 162
    power series, 197
    rational numbers, 226
proof by contradiction, 239
properties of a field, 229

quantifiers, 245

quotient of rational numbers, 227

radius of convergence, 194, 201
range of a function, 258
Ratio Test, 55
    when inconclusive, 56
Ratio Test for Convergence, 53
Ratio Test for Divergence, 55
rational numbers, 1, 225, 226, 235
    as equivalence classes, 226
    construction of, 226
    product of, 226
    quotient of, 227
    sum of, 227
rationals and reals do not alternate, 6
real analysis, 1
real number system, 271
real numbers, 1, 2, 236
    as a subfield of the complex num-
      bers, 14
    constructing, 3, 6
    construction of, 9
    decimal representation, 5
    uncountability of, 5
real numbers are uncountable, 271
real-analytic functions, 194
    elementary operations on, 195, 197
rearrangement of conditionally conver-
    gent series, 65
rearrangement of series, 63
relation, 253, 257
relationship between "for all" and "there
    exists", 247
reversing the limits of integration, 161
Riemann integral, 151, 153
    change of variable in, 162
    elementary properties of, 159
    inequalities about, 162
    linear nature of, 159
    properties of, 158
Riemann sum, 152
Riemann/Weierstrass theorem on re-
    arrangements of series, 63

right limit, 120
Rolle's Theorem, 136
Root Test, 55
    when inconclusive, 56
Root Test for Convergence, 52
Root Test for Divergence, 55
"rule", 253, 254, 258

same cardinality, 262
scalar multiplication of series, 69
Schroeder-Bernstein theorem, 263, 266
sequence
    bounded, 20
    convergence of, 19
    divergence of, 20
    elementary properties of limits of,
      21
    forming a pattern, 19
    non-convergence of, 20
    uniqueness of limit, 20
sequence of functions, 169
    convergence of, 169
sequences, 19
sequences and series
    confusion of, 42
series, 41
    as an infinite sum, 41
    Cauchy criterion for, 44
    convergence of, 42
    divergence of, 42
    important in Fourier analysis, 61
    of functions, 180
    of numbers, 42
    of powers, 52
    operations on, 69
    tail of, 46
    with nonnegative summands, 46
set
    element of, 248
    member of, 248
set-builder notation, 248
set-theoretic
    difference, 252

product, 268
sets, 248
    neither open nor closed, 80
simple discontinuity, 121
sine and cosine
    elementary properties of, 210
sine function, 208
    Taylor series for, 209
smaller cardinality, 262
square root
    existence of, 4
square root of 2 irrational, 1
square root of minus one, 11
square roots
    existence of, 4
subsequence, 27
    convergence of, 27
subset, 249
subtraction
    of integers, 223
    of rational numbers, 228
    of sets, 252
successor, 217
"suffices for", 241
sum of integers, 223
sum of sets, 93
Summation by Parts, 58
summation notation, 41
supremum, 2, 3

tail of a series, 46
Taylor series, 185, 203
Taylor's expansion, 203
Taylor's theorem with remainder, 185
term-by-term integration of power series, 203
"there exists", 245–247
topology, 75
totally disconnected set, 97
transcendental numbers, 68
triangle inequality, 6, 16
trigonometric functions, 206, 208
trigonometric polynomial, 187

"true", 236
truth table, 236, 239

uncountable set, 5, 262, 270, 271
    example of, 271
uniform
    continuity, 115
    convergence, 170
    convergence of a series of functions, 181
    limit of continuous functions is continuous, 171
    limit of integrable functions is integrable, 172
uniform continuity and compact sets, 116
uniform continuity vs. continuity, 115
uniformly Cauchy sequences of functions, 176
union of countable sets, 268
union of sets, 249
uniqueness of limits, 103, 104
universal set, 252
upper bound, 2
    least, 2, 3

value of π, 212
van der Waerden, B. L., 131
Venn diagram, 248, 250, 252

Weierstrass Approximation Theorem, 185
Weierstrass Nowhere Differentiable Function, 131
Weierstrass, K., 185
Weierstrass M-Test, 182
Well Ordering Principle, 267

Zero Test, 45

Printed and bound by CPI Group (UK) Ltd, Croydon, CR0 4YY

21/10/2024

01777085-0010